# NATURAL DISASTERS

# NATURAL DISASTERS

*Editors*

K.K. Singh
Lotfi Aleya
Vinod Singh
Mahadevi Singh

A P H PUBLISHING CORPORATION
4435-36/7, ANSARI ROAD, DARYA GANJ
NEW DELHI-110 002

*Published by*
S.B. Nangia
**A P H Publishing Corporation**
4435-36/7, Ansari Road, Daryaganj
New Delhi 110002
Ph.: 23274050
E-mail : aphbooks@gmail.com

**2026**



*Printed at*
**Balaji Offset**
Navin Shahdara, Delhi 110032

*To*
*Late Janab Tahar Aleya and My Teachers*
*Late Dr. Ishtiaq Ali and Prof. H.S. Rathore,*
*Emeritus Fellow*
*Department of Applied Chemistry,*
*Zakir Husain College of Engineering & Technology,*
*Aligarh Muslim University, Aligarh, India.*

**K.K. Singh**

# Foreword

Disasters are no strangers and occur quite frequently without warning. Disaster has always affected mankind and will probably continue as long as life exists on this planet. This exists all over the world and India is one of the most vulnerable regions. During a disaster, everyone is a victim in one way or another and no one is spared. These events can be frightening for adults, but they cause distress and trauma to children. The increasing incidence of disasters across the globe is creating a devastating impact on the lives, property and livelihoods of people.

According to the United Nations, in 2001 alone, natural disasters of medium to high range caused at least 25,000 deaths around the world, more than double the previous year, and economic losses of around US $36 billion. These figures would be much higher, if the consequences of the many smaller and unrecorded disasters that cause significant losses at the local community level were to be taken into account. The most widely accepted reason for rise in number of natural disasters over the last few decades is due to the continuing growth of the world population as well as rapid industrialization. However, with globalization have come industrialization and its impact on natural environment that has, in turn, led to changing climatic conditions and adverse impact on the environment ultimately leading to a phenomenal surge in natural disasters all over the world.

Globally, natural disasters account for nearly 80 per cent of all disasters affected people. These can occur due to diverse climatic conditions prevalent in different parts of the world. Food security remains one of the main political concerns of climatic change. Reduced water supplies would place additional stress on people, agriculture and environment. However, anthropogenic factors aggravating disasters, ranging from chemical and radioactive pollution to global warming and poverty to human exodus, cannot be overlooked. IPCC (2001) has warned that by the middle of this century, the globe's temperature will rise by anything upto 5.8°C. This climatic change may affect the frequency, timing, intensity and duration of hazardous events such as droughts, floods, ice-cap melt, sea level rise and desertification. These effects will not only further deteriorate the environment and impoverish the quality of life, but also endanger life itself on this planet.

Nearly three million people may have been killed worldwide in past two decades due to natural disasters such as cyclones, earthquakes, floods, etc. Ninety per cent of the natural disasters and ninety five per cent of the natural disasters related deaths worldwide occur in developing countries, in which India has the second largest share. The continent of Asia is particularly more vulnerable to disaster strikes. Each year disasters also account for loss of billions in terms of social and community assets. Apart from direct human costs and destruction, disasters hamper the economic development by decreasing the family income, employment and investible surplus, besides causing outbreak of insect-pests and diseases. The likelihood of a natural occurrence such as earthquakes and negligence resulted accidents are incalculable. Disasters both natural and anthropogenic have negative impact on sustainable development. Speculation may come without warning in spite of several measures being drawn. The need of the hour is, therefore, the integration of scientific, technological, ecological, economic and social dimensions of disaster management with sustainable development. The social, gender and economic inequities should be reduced, if not totally eliminated.

It gives me immense pleasure to introduce the book entitled "Disaster Management". The book which comprises of three volume sets is completely updated and comprehensive account of disaster management. I hope this commendable work will be a unique reference material not only for the teachers and students but also for the administrators, policy makers, planners, NGOs and various user departments/agencies.

I appreciate the efforts made by Dr. K.K. Singh, Prof. Lotfi Aleya, Dr. Vinod Singh and Ms. Mahadevi Singh for compiling and editing the chapters of this book meticulously. Lastly, I would like to congratulate the authors, editors, publisher for bringing out this book.

**(Prof. Claude CONDE)**
Chairman of the University of Franche-comte. Presidence,
1. rue Goudimel, 25000 Besangon, France.

# Preface

In the pre-science era, disasters were looked as punishment or "acts of the divine". With the advent of the 19th century, we started getting more scientific explanations of the causes of the disasters. During the global climate system era (post 1970), with the help of satellites, we started understanding the climate system. Natural disasters are purely natural but climatic change may exacerbate these. Almost a five-fold rise in disasters has been observed over 1960s - 1990s and most of these are of hydro meteorological origins. Floods and droughts are the most frequent and widespread natural disasters resulting in casualties, massive destruction to life and property as well as displacement. The slow onset disaster, drought, has long and deep socio-economic impact on the society. It not only influences the food and fodder production but also result in depletion of water resources leading to deforestation, wind erosion and desertification.

Disasters have been with us for as long as the recorded history and presumably even longer. Since the very beginning, man has been exposed to disasters resulting into injury and loss of life, damage to property and livelihoods. Today's society becomes ever more rapidly vulnerable to natural disasters with burgeoning population, urbanization and other developmental activities. As such man has started misuse, abuse and overuse of natural resources. The natural disasters, such as earthquake, floods, tsunamis, volcanic eruptions, cyclones and landslides occur suddenly with far-reaching, severe physical ecological and socio-economic consequences. They can also occur slowly, like drought and desertification, which are caused by natural phenomena, or by natural phenomena coupled with overt effects of human actions. Global Environment Change has resulted in melting of ice caps, rise in sea level, change in monsoon patterns, droughts and devastating floods. Despite phenomenal progress in agricultural production, the number of undernourished is still on the rise in the developing countries. Socio-economic and political inequities still persist. Challenges posed by hunger and poverty remain as compelling as ever. Disasters, whether natural or man-made, play havoc with the lives of millions of people every year around the globe. Their aftermath is nothing but a grim picture of death, devastation and sufferings. These events can be frightening for adults and traumatic for children. Women, children, old and handicapped suffer most during disaster. Different kinds of disasters impinge different impacts on women. Women are particularly vulnerable because they have little access to resources. Natural and man-made disasters jeopardize economy and development. These catastrophes typically result in the substantial loss of hard earned development gains as well as diverting the development funds towards fuelling disasters relief.

Disasters are no longer viewed as extreme events created by natural forces but as unresolved problems of development. In many areas of the world, disaster losses tend to outweigh the development gains. Thus, the need for an effective disaster management strategy has been acutely felt in many quarters. In the light of recent events throughout the world, ranging from natural disasters such as Orissa super cyclone in 1991 and Gujarat Earthquake in 2001 the Asian Tsunami in 2004 and Hurricane Katrina in New Orleans (US) to the manmade disasters such as the London and Mumbai terrorist attacks, floods in Bihar, the disaster management becomes a key concern. Disasters are phenomenon, which can not be stopped altogether but advance planning can mitigate suffering.

The management of disasters continues to be the major challenge for the government, as it remains the core responsibility of the government. In practice, it is impossible to eliminate all disasters. Although many risks are potentially avoidable, global environmental change and uncertainty about future hazardous events, together with the central role played by human failings, "make the total elimination of environmental disasters, an unrealistic task. For disaster management, whether it is drought, tsunami, earthquake, floods and cyclone, relief is the key instrument for all humanitarian assistance provided by the government as well as donor agencies focused towards the immediate needs of the people for shelter, water and medicines. During IDNDR (1990-1999), enough emphasis was laid on sustainable development and reduction. In developing countries like India, sustainable development and good governance can mitigate natural disasters. While the UN Secretary General, Kofi Annan, in his report on strengthening of the United Nations( paragraph 40) stated that "we need to be better prepared for natural disasters and incorporate disaster risk management into our poverty reduction, development and environmental strategies".

Hence, the approach to the management and prevention of disasters should be linking disasters with poverty alleviation programmes in order to attain the goal of sustainable development. The millennium goal set up by ISDR (2002) during World Summit on Sustainable Development held at Johannesburg was "to intensify our collective efforts to reduce the number and effects of natural and man-made disasters". Thus, an integrated, multi-hazard, inclusive approach to address vulnerability and disaster management, including prevention, mitigation preparedness, response and recovery, is essential for a safer world in the 21st century.

An attempt has, therefore, been made through this book to provide a gist of current and relevant information on various aspects of natural hazards, environmental and technological disasters. The three volumes set contains wide array of topics. The book provides up-to-date comprehensive overview of disaster related problems that mankind faces today. Hope these volumes will serve as a useful reading material for students, researchers, policy makers and all those concerned with the subject. The present book is the outcome of contributions of various renowned scientists/academicians who helped us immensely in compilation of this voluminous book. We are indebted to our esteemed authors for their perseverance towards excellent quality of their work.

We are also thankful to Shri S.B. Nangia, Publisher, A P H Publishing Corporation, New Delhi for excellent production of these Volumes.

**Editors**

## Acknowledgements

A book with a wide range of topics requires specialized inputs. This one is no exception too. It gives us a great pleasure to express our deep sense of gratitude to those who helped us directly or indirectly in compilation of this book. We express our sincere gratitude to Prof. T. Satyanarayana (Department of Microbiology, University of Delhi), Prof. T.N. Singh ( Department of Earth Science, IIT, Mumbai), Dr. V. Phogat, Assoc. Professor (Department of Soil Science, CCSHAU, Hisar), Dr. R.S. Dhillon, Sr. Scientist (Department of Soil Science, CCSHAU, Hisar), Dr. H.K. Malik, Assoc. Professor (Department of Physics, IIT, New Delhi), Dr. Prem Singh Shekhawat, Sr. Lecturer (Department of English, M.S. College for Women, Bikaner), Dr. Jai Bharat Singh, Sr. Lecturer (Department of Geography, Govt. Dungar College, Bikaner), Dr. R.K. Siyag [Project Director (Research), Agriculture & Soil Survey, Krishi Bhawan, Bikaner), Dr. Uday Bhan, Dy. Director [Rajasthan State Seed Corporation (RSSC), Sriganganagar], Sister Dr. Jacinth, Principal (Sophia Sr. Secondary School, Bikaner), Kr. Ripu Daman Singh & Smt. Padam Shree (Sadul Ganj Colony, Bikaner), Kr. Mahendra Singh Rathore & Smt. Alpana Rathore (Vyas Colony, Bikaner ), Shri Vikram Singh & Smt. Poonam Singh (K.K. Colony, Bikaner), Shri. C.P. Vyas (SBI, BSF Branch, Bikaner) for their kind blessings and motivation in this pursuit. We would also like to place on record our sincere thanks to all the eminent scientists/academicians and individuals as well as institutions of India and abroad who supported us immensely by their respective contributions.

We are also thankful to Mr. Govind Vyas (Vyas Computers, Bikaner) for his constant help and taking necessary pains towards finalizing the manuscripts camera ready for printing. Thanks are due to our friends and colleagues Shri Karnail Singh, PA to PD (R), Shri R.S. Bhadauria, JSA, Shri R.P. Singh, Shri R.S. Pandey, Driver, Shri. S.S. Chauhan, Shri Vivek Mittal and Shri Neeraj Bhatnagar.

Miss. Mahima Shekhawat, Miss. Pratibha Shekhawat and Miss. Tanya Rathore deserve special mention for their affection. We are also grateful to our families Manju Singh, Yamna, Anjana, Bhartesh, Harshit, Sami, Selim, Enis, Parikshit as well as our relatives for their patience during the long gestation period of this book.

Finally, we express our gratitude to Shri S.B. Nangia, Publisher (Ashish Publishing Corporation, New Delhi) for publishing this book well in time and in it its present format.

# Contents

# List of Contributors

1. **A.K. Singh,** Department of Physics, Dronacharya College of Engineering, Farrukhnagar, Gurgaon (Haryana)-122001, India.
2. **A.K. Srivastav,** Project Directorate (Research), Agriculture & Soil Survey, Krishi Bhawan, Bikaner (Raj.) - 334001, India.
3. **Agatha Christy,** Sophia Sr. Secondary School, Bikaner (Raj.)-334001, India.
4. **Ali Akram,** Office of the DEO-DPC, Sarva Shiksha Abhiyaan, Bikaner (Raj.)-334001, India.
5. **Alka Tomar,** CMS Environment (Research House), Community Centre, Saket, New Delhi-110017, India.
6. **Anjana Chowdhary,** Department of Botany, Govt. Dungar College, Bikaner (Raj.)-334001, India.
7. **Arun Shairya,** Department of Geography, Dr. B.R. Ambedkar Govt. P.G. College, Sriganganagar (Raj.)- 335001, India.
8. **Asha Sharma,** Lady Elgin Govt. Girls Sr. Secondary School, Bikaner(Raj.)-334005, India.
9. **B.R. Mirdha,** Department of Microbiology, All India Institute of Medical Sciences, New Delhi-110023, India.
10. **Brajeshwari Singh Samant,** Department of Marine Sciences, Berhampur University, Berhampur-760007 (Orissa), India.
11. **D. Venkat Reddy,** Department of Civil Engineering, National Institute of Technology-Karnataka, Surathkal, Srinivasanagar-575025, Mangalore (Karnataka), India.
12. **Durairaj Sambandan,** Department of Zoology, Arignar Anna Government Arts College, Cheyyar-604407 (Tamil Nadu), India.
13. **Eline P. Meulenberg,** ELIT Support VOF, Nijmegen, The Netherlands.

14. **Gayatri Verma,** Department of Chemistry, BSA College, Mathura-211004 (UP), India.

15. **Habib Ayadi,** University of Sfax (Tunisia), Research Unit (00/UR/0907), Plankton and Microbes, BP11713000 Sfax, Tunisia.

16. **H.C. Sharma,** Office of Agricultural Extension, Public Park, Bikaner (Raj.) -334001, India.

17. **Himanshu Sharma,** Army Public School, Bikaner-334001 (India).

18. **Ira Chowdhary,** Sophia Sr. Secondary School, Bikaner (Raj.)-334001, India.

19. **Jani Bergstrom,** Environmental Risk Assessment Centre (ERAC), Geological Survey of Finland, Land Use and Environment, P.O. Box 1237, FI-70211 Kuopio, Finland/University of Kuopio, Faculty of Social Sciences, Department of Health Management and Economics, P.O. Box 1627, 70211 Kuopio, Finland.

20. **Jitender Saroha,** Department of Geography, B.R. Ambedkar College, (University of Delhi), Delhi, India.

21. **K. Murugan,** Division of Entomology, Department of Zoology, School of Life Sciences, Bharathiar University, Coimbatore-641046, (Tamil Nadu), India.

22. **K. K. Singh,** Project Directorate (Research), Agriculture & Soil Survey, Krishi Bhawan, Bikaner (Raj.)-334001, India.

23. **Lotfi Aleya,** Department de biologie Laboratorie de Chrono-Environment CNRS 6249, University of Franche-Comte, CNRS 6249, 1, Place Lecderc, F-25030, Besancon cedex, France.

24. **M.E.A. Mondal,** Department of Geology, Aligarh Muslim University, Aligarh-202002 (UP), India.

25. **M. Raza,** Department of Geology, Aligarh Muslim University, Aligarh-202002 (UP), India.

26. **M. Smita Achary,** Department of Marine Sciences, Berhampur University, Berhampur-760007 (Orissa), India.

27. **M. Stephen Das,** Bikaner Boys School Bikaner (Raj.) - 334001, India.

28. **Mahadevi Singh,** Sophia Sr. Secondary School, Bikaner (Raj.)-334001, India.

29. **Mohammed Alaoui Mhamdi,** Universite Sidi Mohamed Ben Abdallah, Faculte des

Sciences, Departement de Biologic, B. P. 1796, Fes-Atlas, (Maroc).

30. **N. Chitra,** Krishi Vigyan Kendra, National Pulses Research Centre Campus, Tamil Nadu Agricultural University, Vamban, Pudukottai-622303, (Tamil Nadu), India.

31. **Nihal Singh,** Ex-Principal, Nagar Nigam Inter College, Tajganj, Agra (UP).

32. **Nishat Ahmed,** All India Institute of Medical Sciences, New Delhi-110029, India.

33. **Padma, S. Rao,** Air Pollution Control Division, National Environmental Engineering Research Institute (NEERI), Nehru Marg, Nagpur-440020, India.

34. **Poonam Kanwal,** Department of Political Sciences, Janki Devi Memorial College (Delhi University), Sir Gangaram Hospital Marg, New Delhi-110057, India.

35. **R.C. Panigrahy,** Department of Marine Sciences, Berhampur University, Berhampur-760007 (Orissa), India.

36. **R.P. Soundararajan,** Krishi Vigyan Kendra, National Pulses Research Centre Campus, Tamil Nadu Agricultural University, Vamban, Pudukottai-622303, (Tamil Nadu) India.

37. **Rachna Panda,** Department of Marine Sciences, Berhampur University, Berhampur-760007 (Orissa), India.

38. **Rachna Tiwari,** Sophia Sr. Secondary School, Bikaner (Raj.)-334001, India.

39. **Rajender Singh,** Department of Plant Pathology, CCS Haryana Agricultural University, Hisar-125004, (Haryana), India.

40. **Rashmi Rekha Singh,** Freelance Writer, 20, Adarsh Nagar, Agra (UP), India.

41. **Rita K. Bhimaya,** Sophia Sr. Secondary School, Bikaner (Raj.)-334001, India.

42. **S. Devotta,** National Environmental Engineering Research Institute (NEERI), Nehru Marg, Nagpur-440020 (MS), India.

43. **S. Kamalakannan,** Division of Entomology, Department of Zoology, School of Life Sciences, Bharathiar University, Coimbatore-641046 (Tamil Nadu), India.

44. **S.K. Mathur,** Ex-Dy. Director, Quality Control Laboratory, Krishi Bhawan, Bikaner (Raj.)-334001, India.

45. **S.Y. Bodkhe,** Wastewater Technology Division, National Environment Engineering Research Institute (NEERI), Nehru Marg, Nagpur-440020 (MS), India.

46. **Suman Phogat,** Department of Chemistry Polytechnic, Nathusari Chapta, Sirsa-125053 (Haryana), India.

47. **S.S. Verma,** Depatment of Physics, Sant Longowal Institute of Engineering & Technology, Longowal-1418106, Dist. Sangarur, (Punjab), India.

48. **Sunita Chaudhary,** Freelance Writer, C-3/III/2, Civil Lines, Bikaner (Raj.)-334001, India.

49. **Surender Dhankar,** Department of Vegetables, College of Agriculture, CCS Haryana Agricultural University, Hisar-125001, India.

50. **Shakuntala Singh,** Freelance Writer, B-45, Trans Yamuna Colony, Rambagh, Agra (UP), India.

51. **Shera Ram,** Department of Geography, Govt. Lohia College, Churu (Raj.), India

52. **Subodh Kumar Maiti,** Centre of Mining Environment, Indian School of Mines, Dhanbad-826004 (Jharkhand).

53. **Surjit Kaur,** Rashtra Sahayak Sr. Secondary School, Bikaner(Raj.) - 334003, India.

54. **T. Nandy,** Wastewater Technology Division, National Environment Engineering Research Institute (NEERI), Nehru Marg, Nagpur-440020 (MS), India.

55. **V. M. Mhaisalkar,** Environment Engineering, Dept. of Civil Engineering VNIT, Nagpur, India.

56. **Vijay Pal Singh,** Department of Agricultural Zoology & Entomology, RBS College, Bichpuri, Agra (UP)-283105, India.

57. **Vinod Singh,** Department of Geography, Govt. Dungar College, Bikaner (Raj.)-334001, India.

58. **Yogendra Singh,** Department of Biotechnology, College of Agriculture (Rajmata Vijayaraje Scindia Krishi Vishwavidyalaya), Ganj Basoda, Dist. Vidisha (MP), India.

# Editors

**Dr. K.K.Singh:** Dr. K.K. Singh did his B.Sc (Ag.) Hons, M.Sc.(Ag.) in Agricultural Chemistry and Soil Science with first division from Agra University, Agra and M.Phil and Ph.D in Applied Chemistry from AMU, Aligarh. He availed UGC-JRF during his M.Phil, and Ph.D. studies. He is still pursuing P.O. Diploma in Disaster Management from Indira Gandhi Open University, New Delhi. He qualified GATE - 91 examination with an overall percentile score of 91.55. He qualified the National Eligibility Test in Soil Sci. - Soil Chem./Fert. and Microb. in 1995 and NET - 1997 in Soil Sci. -Soil Physics / Soil & Water Conservation organized by Agricultural Scientist Recruitment Board , New Delhi. He joined the Department of Agriculture, Govt. of Rajasthan in 1993 as Asst. Agril. Research Officer (Chem.) in a Gazetted position and is presently working at Project Directorate (Research), Agriculture & Soil Survey, Krishi Bhavan, Bikaner. He has also served as Asst. Professor (Soil & Water Conservation) at Sher-e-Kashmir University of Agricultural Sciences & Technology (RARS, Leh), Srinagar. He has above 100 publications in the form of research papers, book chapters, articles. He has also 10 books published to his credit. He is also a recipient of ARIC 'Crop Research Award-1997' for his significant contribution for promoting literature in the field of agriculture, 20th Century 'Young Agricultural Scientist National Award', and 'Best Paper Award' on Fly Ash. He has attended a number of national/international seminars/ conferences/workshops and presented papers. He was a member of Editorial Board of J. Crop Research from 2001 to 2003. He is Life Member of Indian Science Congress Association. He is also an expert of agricultural programmes broadcast from AIR, Bikaner.

**Prof. Lotfi Aleya:** Prof. Lotfi Aleya did his M.Sc. in Protistology and Ph.D. in Biology of Micro-organisms from Blaise Pascal University (France). He joined the National Center for Scientific Research in Besancon (Franche-Comte University, France) as Professor of Biology and Microbiology. He pilots cooperative research projects in France and other countries, engaging private sector companies and public sector research centers associated with his Chair. He has several research papers and chapters published to his credit. He is responsible for ten of Ph.D. students and is peer reviewer of several scientific journals.

**Dr. Vinod Singh:** Dr. Vinod Singh is M.A. (Geography), M.Phil, and Ph.D. from Jawaharlal Nehru University, New Delhi. His field of interest is environmental and socio-economic problems in urban and rural areas. He is UGC-NET in Human Geography and Population Studies. He has been teaching physical and human geography and climatology to undergraduate as well as post-graduate classes. He is presently working as a Senior Lecturer in Geography at Govt. Dungar College, Bikaner. He has executed a UGC-sponsored project in Urban Wastes Management in Bikaner City. He has several chapters published to his credit and also edited two books. He is actively engaged in supervision of M.Phil., and.P.h.D. scholars on various aspects of geography of the Thar Desert of India.

**Ms. Mahadevi Singh:** Ms.Mahadevi Singh did her M.A. (Sociology) from Dr. Bhim Rao Ambedkar University, Agra and M.A. (English Lit.) from M.D.S. University, Ajmer. She is still pursuing M.Phil in English from Madurai Kamraj University, Madurai. She has a long teaching experience and has 21 publications to her credit in form of articles and book chapters. She has a vast expertise on social and gender issues. She has also delivered radio talks on women issues from AIR, Bikaner.

# Section I

## EARTHQUAKES, LANDSLIDES FLOODS AND DROUGHT

## Chapter 1

# Disaster Management : An Overview

*K.K. Singh*[1] *and Mahadevi Singh*[2]

## INTRODUCTION

Since the beginning of the human civilization, we have been facing fury of the nature like earthquakes, floods, droughts, cyclones, etc. The world is becoming increasingly vulnerable to natural disasters. The increasing incidence of disasters across the globe is creating a devastating impact on the lives, property and livelihoods of people. India, on account of its geographical position, climate and geological setting is one of the worst affected areas of natural disasters like floods, cyclones, drought, earthquakes, forest fires, landslides and tsunami in the south Asian region (Singh and Mahto, 2007).

The natural disasters, such as earthquakes, floods, tsunamis, volcanic eruptions, cyclones and landslides occur suddenly with far-reaching, severe physical, ecological and socio-economic consequences. They can also occur slowly, like drought and desertification which are caused by natural phenomena, or by natural phenomena coupled with overt effects of human actions (Santra, 2001). Natural disasters are of three types: (i) The disasters caused by extreme geotechnical events; (ii) The disasters caused by extreme weather conditions; and (iii) The disasters caused by other extreme incidents (Frater, 1998). Earthquakes, tsunamis, volcanic eruptions, avalanches, and landslides occur due to geotechnical events; flooding, sea floods, hurricanes/tornadoes, hailstorms occur due to weather conditions; and forest fires, desertification, steppe formation, severe droughts occur due to other extreme events such as reduction in rainfall.

Evidence suggests that the frequency of natural disasters has increased in recent years. The disasters occurred three times higher

[1]Project Directorate (Research ), Agriculture & Soil Survey, Krishi Bhawan, Bikaner (Raj.) -334001, India, Email: singhkk95@yahoo.com

[2]Sophia Sr. Secondary School, Bikaner (Raj.) - 334001, India.

worldwide in the past ten years than in the 1960s and economic losses were eight times greater, exceeding US $ 60 billion a year. For the three years 1993-95, the United States of America Sub-Committee on Natural Disaster Reduction estimated that annual losses from natural disasters in the country averaged one billion dollars per week (Sinha, 2001). The most widely accepted reason for the rise in number of natural disasters over the last few decades is due to the concentration of population in mega-cities. In addition, changes in the global environment threaten us with the possibility of severe cyclones, rising sea levels, droughts, among others. Large scale migration of populations throughout the world to coastal regions that are generally more vulnerable to natural disasters like tsunamis has contributed to increased loss of life and property.

The impact of disaster does not depend only on the force with which it strikes, but the way it is received and felt is equally important (Sinha, 2001). Low - income societies are the worst sufferers of the disasters and the poor people are the worst hit. Most of the disaster problems are essentially the unsolved development problems, especially in the developing countries. Therefore, disaster prevention measures need to be implemented as important basics of development. Hence, there is a need for a holistic approach to the management of disasters on the basis of linkages between disaster and poverty, environment and sustainable development.

In this article, an attempt has therefore, been made to discuss types of disasters, their impacts, along with management.

## CONCEPT AND MEANING OF DISASTERS

The term "Disaster" owes its origin to the French word "desastre" which is a combination of the two words 'des' meaning ‘bad’ and 'astre' meaning ‘star’. Thus, the term "disaster" refers to “bad" or “evil star". National Disaster Management Act 2005 of India defines a number of terms related to disasters and these will naturally prevail in India. The terms defined in the Act, *interalia,* include disaster, disaster management, mitigation, and preparedness. There are however, terms related to disasters other than those in the Act which will also need defining towards establishing a common vocabulary across the country.

Some of the terms defined in the Act are not simple and may leave a wide margin for interpretation while in use, take for instance definition of a disaster. The National Disaster Management Act 2005 defines disaster as “a catastrophe, mishap, calamity or grave occurrence affecting any area, arising from natural, or man-made causes or by accident or negligence, which results in substantial loss of life or human suffering or damage to, and destruction of, property or damage to, or degradation of environment, and is of such a nature or magnitude as beyond the coping capacity of the community of the affected area." It remains to be seen as how are we going to interpret an event when terms like catastrophe, mishap, calamity and grave occurrence co-exist and will naturally be seen as interchangeable. And if to complicate the matters further, the definition of hangs on the knife edge of local coping capacity. Suppose there are two identical events at two different locations A and B, and the local coping at A is higher than that required to manage the event whereas at B the local coping capacity is much lower. Clearly it -

UNDRO-UNDP Training Programmer - An Overview of Disaster Management, 1992, second edition; would mean that the same event would be recognized as a disaster at location B but not at location A?

UNDRO Disaster Management Training Manual 11 defines "Disaster as a serious disruption of the functioning of a society, causing widespread human, material, or environmental losses which exceed the ability of the affected society to cope using its own resources."

The United Nations (UNDRO, 1987, cited in Hanisch, 1996, p. 22) defines disasters in the following way. "A disaster is an event that is concentrated in space and time and that subjects a society to severe danger and such serious losses of human life or such major material damage that the local social structure breaks down and the society is unable to perform any or some of its key functions."

The High Powered Committee of the Government of India, in its October 2001 Report 12 defines Disaster as "an occurrence of a severity and magnitude that normally results in deaths, injuries, and property damage and that can not be managed through the routine procedures and resources of government. It usually develops suddenly and unexpectedly and requires immediate, coordinated and effective response by multiple government and private sector organizations to meet human needs and speedy recovery."

The definition of disaster provided by the Center for Research on the Epidemiology of Disasters (CRED) is relatively simpler. CRED defines a disaster as a situation or event which overwhelms local capacity, necessitating a request to national or international level for external assistance; an unforeseen and often sudden event that causes great damage, destruction and human sufferings. Then it goes on to add that for a disaster to be entered into the database, at least one of the following criteria must be fulfilled : (i) ten or more people recorded killed, (ii) 100 people reported affected, (iii) declaration of state of emergency, and (iv) call for international assistance.

Clearly, the difficulty of the kind faced in the definition given in the Act does not arise if such a definition is used.

## COMPONENTS OF A DISASTER

Natural events only become potential hazards when they threaten people or property. An earthquake will cause little damage if it takes place in an empty desert. It may also cause little damage if it takes place in a city like San Francisco, where people can afford to be well protected. A natural event only causes serious damage when it affects an area where the people are at risk and poorly protected. Disasters occur when these two factors are brought together:

- People living in unsafe conditions,
- Natural hazards such as flood, hurricane or earthquake.

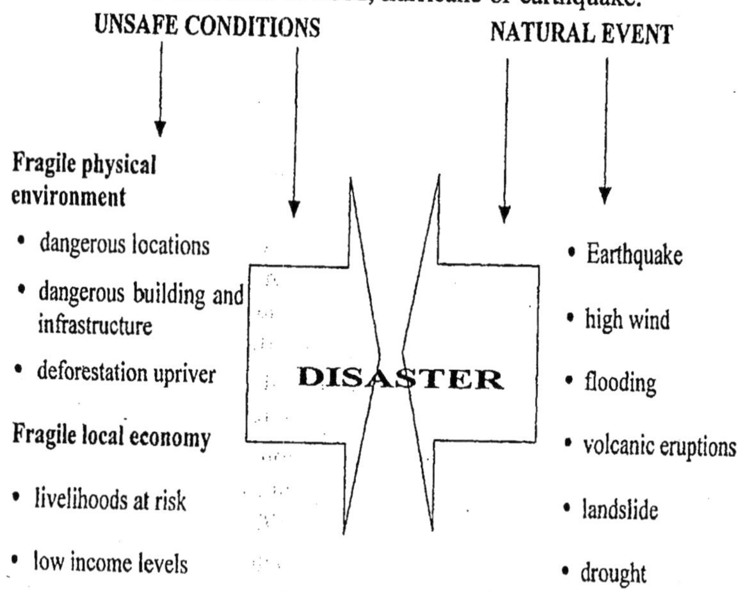


The natural hazard is often blamed for the disaster but, in fact, the real cause may be that the people were poor and unprotected. Many poor people know that they are living in areas with a high risk of, for example, regular flooding or earthquakes.

Often they simply can not afford to live or elsewhere. They have no choice but to take these risks.

Understanding the real causes of disasters help us to realize how situations could be improved through appropriate support and development programs. Balanced community development has resulted in a protected environment with stronger housing and buildings and a healthy local economy. There are also protection measures such as windbreaks, flood control measures and an early warning system which provides at least 24 hours warning of likely cyclones, hurricanes, earthquakes, etc.

No measures can ever fully protect against all possibilities but these ideas, if put into practice, would bring huge benefits.

**What is disaster ?**

A disaster is ...

- A sudden calamitous event bringing great damage, loss or destruction (Merriam Webster dictionary);
- Situation or event which overwhelms local capacity, necessitating a request to national or international level for external assistance;
- An unforeseen and often sudden event that causes great damage, destruction and human sufferings;
- An event, concentrated in time and space, which threatens a society or a relatively self - sufficient subdivision of a society with major unwanted consequences as a result of precautions which had hitherto been culturally accepted as unwanted (Turner, 1976);
- Some rapid, instantaneous or profound impact of the natural environment upon the socio-economic system (Alexander, 1993);
- An extreme event as any manifestation of the earth's system (lithosphere, hydrosphere, biosphere or atmosphere ) which differs substantially from the mean (Alexander, 1993);
- The event which becomes a disaster when the community's capacity to cope is overwhelmed and the status quo becomes untenable (Akkihal, 2006 );
- A disaster is defined as a serious disruption of the functioning of society, causing widespread human, material or environmental losses

that exceed the ability of affected society to cope using only its own resources [(UNDHA, 1992) cited in Russell, 2005];

- The term 'natural disaster' is commonly used when describing the impact of a natural hazard on a community (Russell, 2005);
- They are the outcomes of human settlement patterns, land use decisions, and the use of risky technologies [Russell, 2005];
- An event that results in death or injury to humans, and damage or loss of valuable goods, such as buildings, communication systems, agricultural land, forest, natural environment, etc.

**Definition of Disaster**

A serious disruption of the functioning of a community or a society causing widespread human, material, economic or environmental losses which exceed the ability of the affected community or society to cope using its own resources.

**Classification**

Disasters can be categorized in various ways but based on the origin, hazards worldwide are basically grouped in two broad groups:

1. Natural hazards, i.e., hazards with meteorological, geological or even biological origin.
2. Manmade hazards, i.e., hazards with human - induced or technological origin.

Natural hazards and disasters fall into two categories, e.g., (i) planetary hazards/disasters, and (ii) extra - planetary or extra-terrestrial natural hazards/disasters. The natural and man-made hazards are classified in Fig. 1.

Hazards can be of both *short-term* and *long-term* duration, as per the classification proposed by Smith (1996). Hazards and the disasters they cause are classified as *"rapid on set or cataclysmic,* and long-term or continuing". In a cataclysmic disaster, one large-scale event causes most of the damage and destruction but soon dissipates. Recovery in the sense of restoration of equilibrium conditions is relatively faster. In a long-term, continuing disaster, the situation following the event remains constant or may even deteriorate further. Examples of cataclysmic disasters are earthquakes, volcanic eruptions, cyclonic storms and floods. Examples of continuing natural disasters are droughts, crop failures, environmental degradation such as deforestation and desertification, etc. The damaged

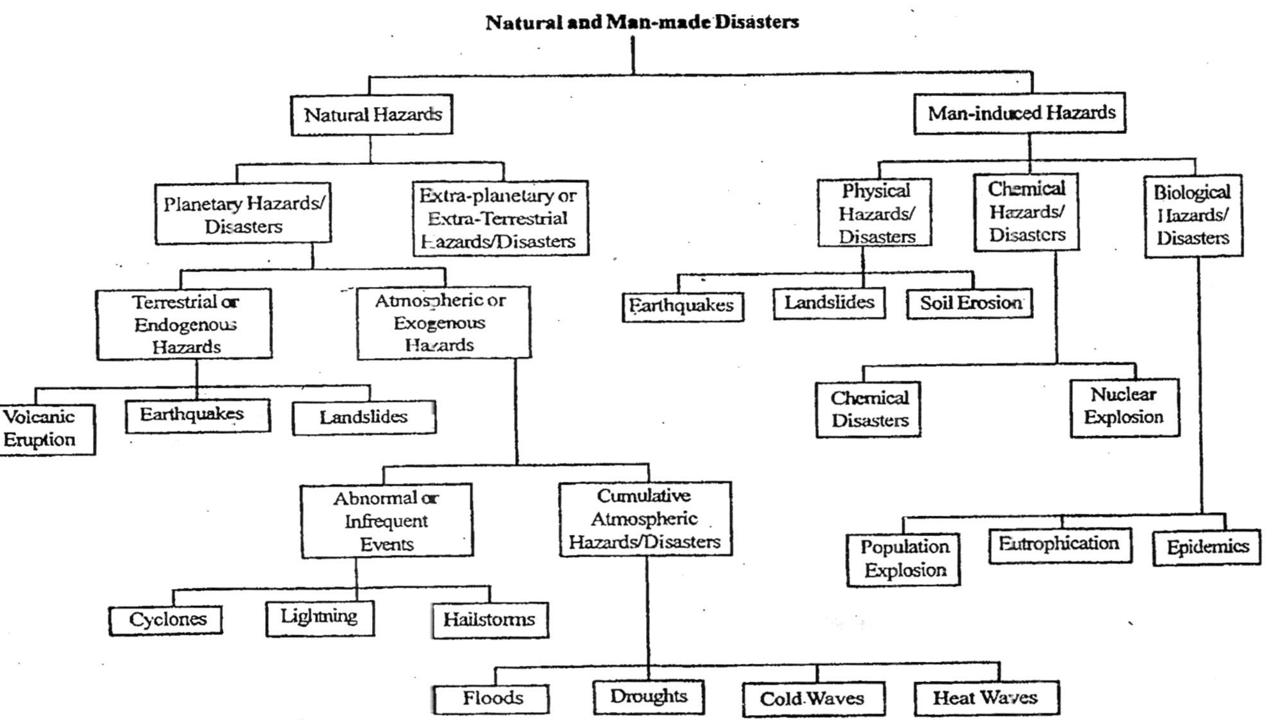


**Fig. 1 : Classification of Natural and Man-made Disasters**

area in a cataclysmic disaster is usually relatively small while the area affected in a continuing disaster stretches over a wide expanse, spatially and temporally long-term impact which affects the past, present and future of communities.

Disasters are also classified as compound and complex . A compound disaster is explained as one crisis leading to other contingences, such as famines followed by civil strife, mass displacement of people, etc. Complex disasters are those that lead to collapse of the political authority or lead to some other complexity where the problems involved/generated are intensely political in nature, such as communal bias in distribution of relief, relocation of communities, compensation, disbursement, etc.

Disasters identified by the High Power Committee are:

***I. Water and climate related disasters***

1. Floods and Drainage Management
2. Cyclones
3. Tornadoes and Hurricanes
4. Hailstorm
5. Cloud Burst
6. Heat and Cold Wave
7. Snow Avalanches
8. Droughts
9. Coastal Erosion
10. Thunder and Lightening

***II. Geological disasters***

1. Landslides and mudflows
2. Earthquakes
3. Dam Failures/Dam Bursts
4. Mine Fires

***III. Chemical, Industrial and Nuclear related disasters***

1. Chemical and Industrial Disasters
2. Nuclear Disasters

***IV. Accident related disasters***

1. Forest Fires
2. Urban Fires

3. Mine Flooding
4. Oil Spill
5. Major Building collapse
6. Serial Bomb Blasts
7. Festival Disasters and Building Ablaze fire
8. Electrical Disasters and Fires
9. Air, Road and Rail Accidents
10. Boat capsizing
11. Village Fires

***V. Biological disasters***

1. Biological disaster and Epidemics
2. Pest Attacks
3. Cattle Epidemics
4. Food poisoning

**Disasters : World Scenario**

Nearly three million people worldwide may have been killed in the past 20 years due to natural disasters such as landslides, earthquakes, floods, snow avalanches, cyclones, etc. Ninety per cent of the natural disasters and ninety five per cent of the natural disasters related deaths worldwide occur in developing countries in which India has the second largest share (Singh and Mahto, 2007).

According to the United Nations, in 2001 alone, natural disasters of medium to high range caused at least 25,000 deaths around the globe, more than double the previous year, and economic losses of around US $ 36 billion. These figures would be much higher, if the consequences of many smaller and unrecorded disasters that cause significant losses at the local community level are to be taken into account. Devastations in the aftermath of powerful earthquakes that struck Gujarat, El Salvador and Peru; floods that ravaged many countries in Africa, Asia and elsewhere; droughts that plagued Central Asia including Afghanistan, Africa and Central America; the cyclone in Madagascar and Orissa; Bubonic plague spread by fleas on rat - plague epidemics in Asia, Africa and South America and floods in Bolivia are global events. On December 26, 2003, nature had also reminded us of her presence. The historic city of Bam, in Iran has been all but reduced to rubble. In recent memory, Messina in Italy,

San Francisco in USA, Kobe in Japan, Asian Tsunami, in Kashmir in India and Pakistan, Siachun in China have all been devastated by massive earthquakes. Moreover, terrorist attacks upon human, targets such as World Trade Centre on September 11, 2001 and terrorist attacks on in Mumbai on November 26, 2008 are the culmination of heinous act of terrorists.

Storm-related disasters have also been increasing in frequency and intensity during the past decade. The high winds that accompany tropical storms and the resulting floods have a particularly devastating impact on agriculture. The most recent world Disaster Report of the International Federation of the Red Cross shows that during 1990-1999 wind storms and flood-related disasters together accounted for 60% of the total economic loss caused by natural disasters. A significant percentage of disaster causalities, in terms of deaths, injuries and people displaced from the homes and livelihoods, is also attributable to storms and floods. The World Watch Institute estimated that during the first 11 months of 1998, weather - related disasters caused more than $ 89 billion in economic losses, resulted in 32,000 deaths and displaced 300 million people from their homes and livelihood systems. Most of the disasters in 1998, including record floods in China and Bangladesh and Hurricane Mitch in Central America, were attributed to the *El Nino* or *La Nina* phenomena. The direct and indirect economic cost of the floods in Mozambique caused by tropical storms Elyna and Gloria in February and March 2000 is estimated at US $ 1 billion.

Floods in more than 80 countries have caused hardships for more than 17 million people worldwide since the beginning of 2002, according to the World Meteorological Organization (WMO). Almost 3,000 people have lost their lives while property damage is amounting to over thirty billion US dollars. The total area affected by the floods is over 8 million $km^2$, almost the size of the United States of America.

Recently, Nargis cyclone and Sichuan earthquake have taken place. Cyclone *Nargis* was a tropical cyclone with a speed of 165 $kmh^{-1}$ that caused the worst natural disaster in the recorded history of Myanmar on May 2, 2008 (The Australian, 2008). The fatalities were at least 90,000 in number with a further 56,000 people still missing (BBC News, 2008). The 2008 Siachuan earthquake or Great Siachuan Earthquake, which measured at 7.9 on Richter scale according to USGS, occurred on $12^{th}$ May, 2008 in Sichuan province of China. The death toll was reported as 69,163 confirmed and 374,142 injured with 17,445 listed as missing. The earthquake left about 4.8 million people homeless (Ramzy, 2008 and Hooker, 2008).

However, what is disturbing is the knowledge that these trends of distribution and devastation are on the rise instead of being kept in check. At the global level there has been considerable concern over natural disasters. Even as substantial scientific and material progress is made, the loss of lives and property due to disasters has not decreased. In fact, the human toll and economic losses have mounted. It was in this background that the United Nations General Assembly in 1989, declared the decade 1990-2000 as the "International Decade for Natural Disaster Reduction" with the objective to reduce loss of lives and property and restrict socio-economic damage through concerted international action, specially in developing countries.

**Disasters : Indian Scenario**

India supports 1/6$^{th}$ of the world's population on just 2% of its landmass. It suffers heavily from natural disasters of every shade and description that hits the poorest of the poor. According to one estimate, nearly 59% of India's land area is prone to earthquakes of moderate to high hazard, nearly 12% is flood prone, about 8% is cyclone prone, 2% is prone to landslides and a long coastline is exposed to tsunamis and storm surges. Drought, regarded as disaster in slow motion, affects as much as 68% of India's land area. Of the 35 states and Union Territories, as many as 27 are disaster prone. And if the perceived threats due to other disasters such as chemical and terrorist attacks are added, every square inch of India is vulnerable, calling for immediate attention and sustained effort.

Losses due to disasters both direct and indirect, elude reliable estimates. According to one World Bank estimate, reported direct losses are of the order of $ 30 billion over the past 35 years. During the period 1994-98, approximately 120 million people were affected by natural disasters in one-way or the other and according to one estimate economic losses piled to about 28,000 crore. During the period 1998-2003, the losses amounted to Rs. 47000 crore. In 2005 also, disasters in India caused direct losses approaching Rs. 87,500 crore.

Moreover, India is experiencing massive and rapid urbanization. It is estimated that by 2025 AD, the urban component, which was only 25.7 per cent in 1991 will be more than 50 per cent. The 9$^{th}$ Five Year Plan estimates India's population size by 2011 to be 1179.89 million with an urban population share of 32 per cent. Urbanization is increasing risk at unprecedented levels. Communities become increasingly vulnerable when high density areas with poorly built and maintained infrastructure are subject to natural hazard. Urbanization dramatically increases vulnerability whereby communities are forced to squat on environmentally unstable areas such as steep hillsides prone to landslides along the sides of rivers that regularly flood, or on poor quality ground causing building collapse (Sharma, 2002).

**Distribution and Dimensions of Disasters**

Most of the world's natural disasters occur in countries in Asia and the Pacific, causing enormous destruction and human sufferings. The unique geographical position of the Indian sub-continent makes this region highly vulnerable to natural disasters like drought, floods, cyclones, tsunamis, landslides, and earthquakes, etc. It is estimated that almost 3 million people have perished as a result of natural disasters in the past three decades while tens of millions suffered hardships (UN, 1997). Things appear to be getting worse in two ways : natural disasters appear to be becoming more frequent and their effects more severe. The data pertaining to the impact of disasters in the different continents are presented in table 1. The continent of Asia is the worst -hit in terms of number of human lives lost, number of people affected and disasters related damages. It is important to note that hazards are striking even those area, which are not hazard - prone. Direct monetary losses in terms of damage to transportation systems, communication facilities, electricity networks, loss of crops, loss of livestock, etc. are phenomenal and increasing with time. Obviously, this increase in damages due to disasters is being brought about by concentration of people in disaster - prone areas in spite of repeated warnings to vacate the areas.

In developing countries like India, there has been a vicious cycle of poverty, environmental degradation and disaster. Over - exploitation of fuel wood, extensive cultivation and excessive grazing often cause deforestation and soil erosion. Soil erosion ultimately leads to reduction in soil productivity and increased desertification if preventive measures are not taken. Soil erosion also causes problems of heavy sedimentation/ siltation in the downstream that aggravates flood havoc. A typical example of this kind may be found in Nepal - India - Bangladesh region. It should be noted that natural disasters do not respect political boundaries. Figure 1 shows the number of disasters that occurred across the globe during the last decade. The continent of Asia is the worst - hit (74.20%) when compared to the world in the number of disaster events during the last decade. On the other hand, the percentage of people killed in Asia alone were 74.20% during 2006. It clearly indicates that disasters are increasing in magnitude and intensity in recent years.

**Table 1: Damages due to Natural disasters during 1990-1999**

| *Items* | *Africa* | *America* | *Asia* | *Europe* | *Oceania* | *Total* |
|---|---|---|---|---|---|---|
| No. of human lives lost | 85,551 | 81,058 | 406,199 | 16,115 | 3,614 | 592,537 |
| No. of people affected (millions) | 10.4 | 5.7 | 176.0 | 1.9 | 1.9 | 195.9 |
| Estimated damages (billion US $) | 25 | 178 | 401 | 148 | 11 | 741 |

**Source:** Sinha, 2001

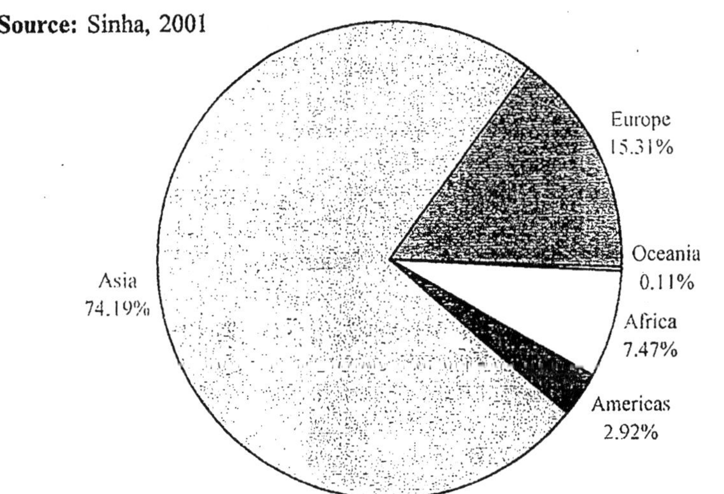


**Fig. 2 : Percentage of people killed by natural disasters by continent**

**Source:** The OFDA/CRED International Disaster Database, Universities Catholique de Louvain, Brussels, Belgium ( www.em_dat.net)

India, characterized by the unique geodynamics, typical monsoon behaviour, flood prone river basins coexisting with semi-arid and arid regions and long coastlines, is amongst the nations most vulnerable to natural disasters. Floods, droughts, cyclones, earthquakes and landslides are the recurring natural disasters in the country. About 60% of the landmass is prone to earthquakes of various intensities; over 40 million hectares is prone to floods; about 8% of the total area is prone to cyclones and 60% of the area is susceptible to drought. Thus, the vulnerability of India to all types of disasters is well known. Some of the major disasters that affect the vast areas of India are dealt with as follows :

### 1. Earthquakes

When fracturing of rocks in the earth's crust takes place due to geological reasons, a series of shock waves travel outward in all directions resulting in rapid and discernible tremors or earth movements. Earthquakes are considered to be one of the most dangerous and destructive natural hazards. Their convulsions have so often reduced man to helplessness that he has always feared them (Bruce, 1978). The impact of this phenomenon is sudden with very little or without any warning, making it just impossible to predict its occurrence and magnitude. A very large number of earthquakes occur every year all round the earth but only a limited number of them are centered near populated areas.

India is no stranger to earthquakes - moderate or great while there is no part of India that could be considered 'aseismic', i.e., free from seismic activity. About 56 per cent of total area of the country is vulnerable to seismic activity of varying intensities (Gaur, 2003). Most of the vulnerable areas are generally located in Himalayan and sub-Himalayan regions, the Andaman & Nicobar Islands, Kachchh and the North-eastern states. India has been facing devastating earthquakes on a regular basis causing widespread damages. At least six very great earthquakes of magnitude 8+ occurred in the past 200 years but earthquakes of less magnitude are also known to have caused huge devastation in terms of destruction of buildings; structures and loss of lives. It is because the nature and extent of the impact of an earthquake depends upon a number of factors such as its magnitude and local depth, geological and soil condition of the region around the epicenter, density of population, extent and types of constructions, time of occurrence whether at night or during the day and any other aggravating situation like fire or flood resulting from it (Jai Krishna, 1992; Tandon, 1992).

On the basis of magnitude of damage risk, India is divided into five damage risk zones :

1. **Zone I of least damage risk :** Some parts of Punjab and Haryana, plains of U.P., portions of plains of North Bihar and West Bengal, delta area of coastal plains of Godavari, coastal plain areas of

Maharashtra and Kerala, desert areas of Rajasthan and most areas of Gujarat except Kachchh region.

2. **Zone II of low damage risk** : Southern Punjab and Haryana, Southern parts of plains of Uttar Pradesh, Eastern Rajasthan, Coastal districts of Orissa, Tamil Nadu, etc.

3. **Zone III of high damage risk** : Areas of Southern and South-eastern Rajasthan, most of Madhya Pradesh, Maharashtra and Karnataka, Southern Bihar and North-western Orissa.

4. **Zone IV of high damage risk** : Jammu & Kashmir, Himachal Pradesh, northern Punjab and Haryana, Delhi, western Uttar Pradesh, Tarai and Bhabar regions and Himalayan region of Uttaranchal and Bihar, and Sikkim areas.

5. **Zone V of very high damage risk** : Parts of Jammu & Kashmir, some parts of Himachal Pradesh, Uttaranchal, extreme North Bihar, entire North-eastern India and Kachchh region of Gujarat.

Though the plains of West Bengal come under the zone of least damage risk but the devastating severe earthquake of Kolkata on October, 1737 killing 300,000 people put a question mark against this concept. The zone of very high damage risk of Kachchh region of Gujarat registered most devastating killer earthquake on January 26, 2001 (8.1 on Richter scale) in its seismic history of past 180 year killing 50,000 to 100,000 people. The towns of Bhuj, Anjar and Bhachau were flattened and razed to the ground. A map of disaster prone areas is shown in Fig. 3.

The earthquakes in India along the Himalayas and foothills may be explained in terms of plate tectonics. The Asiatic plate is moving southward whereas the Indian plate is being subducted below the Asiatic plate. This collusion of Asiatic and Indian plates and subduction of Indian plate and consequent folding and faulting and gradual rise of the Himalayas at the rate of 5 cm per year cause earthquakes of northern India, Tibet and Nepal. According to Negi *et al.* (1988) the Indian sub-continent has deformed at places due to the Indian Ocean floor spreading process. India folds at places and when the energy reaches the elastic limit, the rocks break up and trigger strike - slip and thrust fault earthquakes. These tectonic movements cause earthquakes in northern and North-eastern parts of India.

Table 2 lists the noteworthy earthquakes that occurred in India in the recent past.

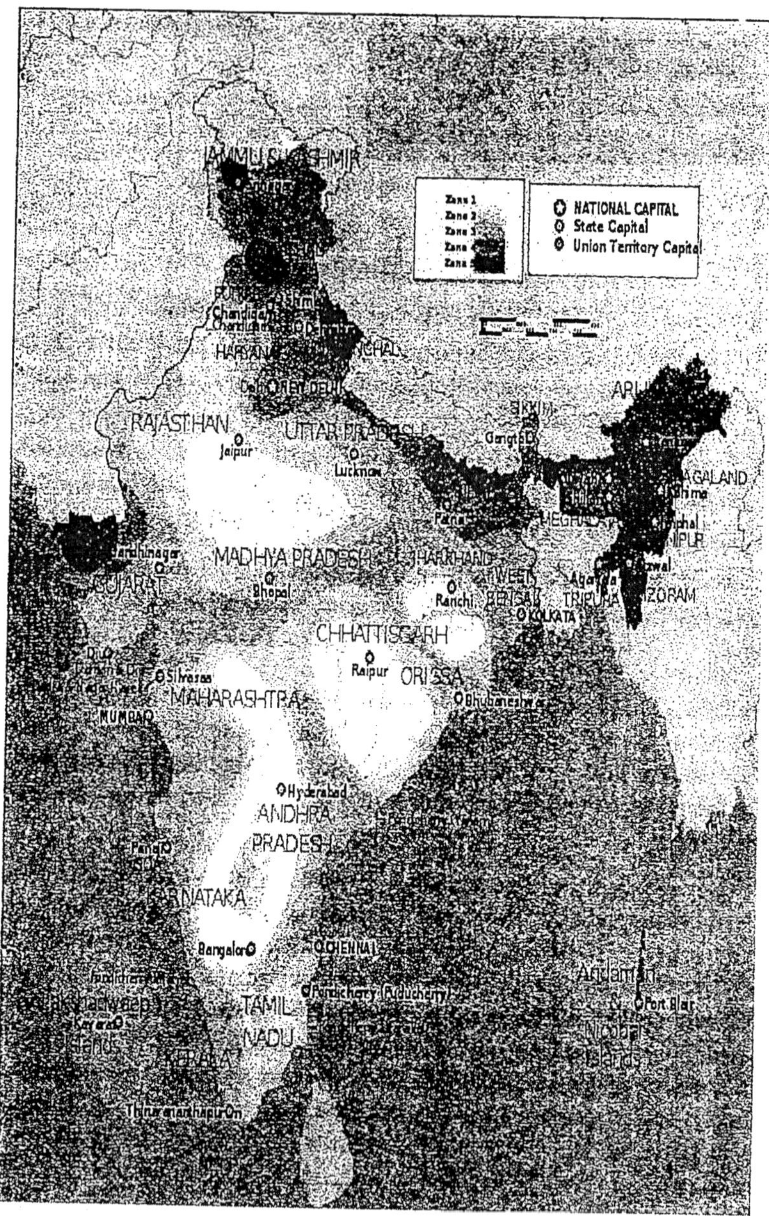

**Fig. 3: Earthquake damage risk zone of India**

**Table 2: Major earthquake events in India**

| *Place* | *Year* | *Magnitude (Richter Scale)* | *Remarks* |
|---|---|---|---|
| Uttarkashi | 1991 | 6.6 | About 750 people died and lots of damage to roads and property. |
| Latur | 1993 | 6.4 | About 8,000 people died huge devastation in Maharashtra and more particularly Marathwada region. |
| Jabalpur | 1997 | 6.0 | About 39 people died and huge damage happened in rural areas. |
| Chamoli | 1999 | 6.8 | About 100 people lost their lives. Huge damage to property took place. |
| Bhuj | 2001 | 6.9 | Devastating earthquake with death toll of 14,000 people. |
| Coastal area | 2003 | 9.0 | Death toll of 1 1,628 lives (Tsunami) |
| Kashmir -Muzaffarabad | 2005 | 7.6 | Devastating earthquake with more than 1,00,000 lives. |

## 2. Tsunami

It is a Japanese word meaning "harbor wave". These waves, which often affect distant shores, originate from undersea or coastal seismic activity, landslides and volcanic eruptions. Whatever the cause, sea water is displaced with a violent motion and swells up, ultimately surging over land with great destructive power. It has claimed over 1,50,000 lives in South Asia.

On 26th December, 2004, a massive earthquake of magnitude 9.0 hit Indonesia generating Tsunami waves in South-east Asia and Eastern coast of India. Height of tsunami waves ranged from 3 to 10 m affecting a total coastal length of 2260 km in the states of Andhra Pradesh, Tamil Nadu, Kerala and Union Territories (UTs) of Pondicherry, Andaman & Nicobar Islands. Tsunami waves traveled upto a distance of 3 km from the coast killing more than 10,000 people and affected more than lakh of houses leaving behind a huge trail of destruction (Anonymous, 2005).

The South Asian Tsunami on 26th December, 2004 left behind a great tragedy in the lives of children causing tremendous loss in their life. Statistics brings out that 37-39 per cent of total deaths were children below the age of 14 years. Many children became orphans and lost one of their parents, their houses and familiar environment. Normal process of life was disrupted creating a sense of panic and confusion in children hampering the normal process of life.

The December 2004 Tsunami and its aftermath in terms of its impact on the lives and livelihood of the coastal communities in Tamil Nadu and Pondicherry have alerted us to compelling issues concerning disaster preparedness and management.

Tsunami is not a single giant wave. It consists of ten or more waves which is termed as a "tsunami wave train". Since scientists can not predict when earthquakes will occur they can not predict exactly when a tsunami will be generated. With the use of satellite technology, it is possible to provide nearly immediate warnings of potentially tsunamigenic earthquakes. The warning includes predicted times at selected coastal communities where the tsunami could travel in few hours.

**3. Floods**

Everyone knows that generally speaking, flood means inundation due to overflowing of a large volume of water, whatever be its source. According to WMO (1974) flood is a rise, usually brief, in the water level in a stream to a peak from which the water level recedes at a slower rate. On the other hand, flood or river flood is a relatively high flow or stage in a river, markedly higher than the usual. A mass of water rising, swelling and overflowing land - ICID Multilingual Technical Dictionary of Irrigation of Drainage (ICID, 1996).

Two facts emerge from these definitions of flood. Firstly, flood is a phenomenon created by water. There is no concept of flood without large and unmanageable volume of water. Secondly, flood in most cases, is a phenomenon of streams and rivers and their surrounding areas.

However, areas with no stream or river may also suffer from floods in the case of inadequate drainage - a situation being experienced increasingly by unplanned growth in urban centres in addition to that created by blockages in agricultural areas. This is called "flooding" or "inundation" to distinguish it from river floods, although both the words, "floods" and "flooding", are often used synonymously. Areas away from river or canal banks and not usually prone to floods may experience flooding if there is a sudden rush of water in river or canal. In addition, there are large bodies of water, besides streams and rivers, which could spill over to create flooding. For example, Himachal Pradesh suffered heavy floods from the spill from the Parichu lake in 2005 and earlier as well.

Among all the disasters that occur in the country, river floods are the most frequent and often the most devastating. The scale of many floods is often extensive, and vulnerable populations and the built environment can be particularly susceptible to the impact. Moreover, the current prediction of continuing climate change is likely to exacerbate the situation (Anjaneyulu, 2004).

River flood, the most devastating disaster in India is mainly caused due to the variations in rainfall, climate and topography. Out of the total rainfall in the country, 75% is concentrated over a short monsoon season of three to four months. As a result, there is a very heavy discharge from the rivers during this period causing widespread floods. As much as 40 million hectare of land in the country has been identified as flood prone and an average of 18.6 million hectare of land is flooded annually. Annually flood affected area, flood damages and people affected are increasing with the flood magnitudes, varying from year to year and region to region. The northern region of India is more vulnerable to floods. The various river systems and their catchments are shown in Fig. 4.

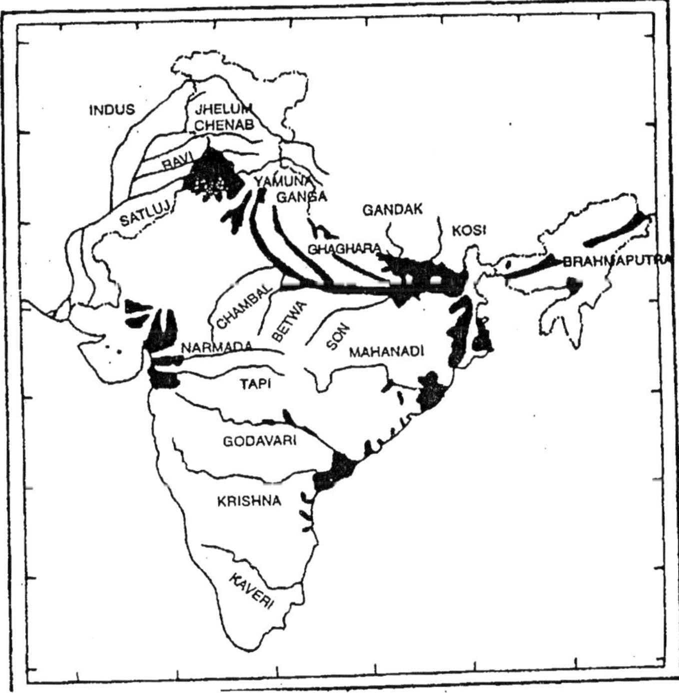


**Fig. 4 : Flood prone area of India**

According to the Central Water Commission (CWC), a total of 14.37 million hectares of land area has been protected in various states. Floods are caused mainly in the Ganga - Brahmaputra - Meghna basin, which carry 60% of the nation's total river flow. India is facing severe damages due to floods on regular basis. A summary of flood damages is given in table 3.

**Table 3 : Annual average flood damages in India (1953-94)**

| *S. N.* | *Items* | *Unit* | *Number* |
|---|---|---|---|
| 1. | Area affected | Million ha | 7.56 |
| 2. | Population affected | Million | 32.03 |
| 3. | Human live lost | Number | 1504 |
| 4. | Livestock/Cattle lost | Number | 96713 |
| 5. | Houses damaged | Rs. (Crores) | 11683.04 |
| 6. | Value of damaged houses | Rs. (Crores) | 136.615 |
| 7. | Crop damaged | Rs. (Crores) | 460.07 |
| 8. | Public utilities damaged | Rs. (Crores) | 377.248 |
| | **Total Losses** | **Rs. (Crores)** | **982.126** |

Though disasters can not be stopped, it is possible to minimize the flood damages with efficient flood disaster management by application of modern technology tools. Space technology plays a potential role in flood disaster management in the forecasting, warning, relief, rescue, rehabilitation and mitigation phases. Presently, CWC has its flood forecasting and warning network, which covers 62 major inter state river sub-catchments that include 132 water level forecasting stations and 25 inflow forecasting stations for important reservoirs. Based on the hydrological and hydro-meteorological data collected from about 700 stations, the CWC issues flood forecasts and flood warning messages generally 24 and 48 hours in advance.

### 3. Droughts

Droughts are a recurring feature in some states of India. Approximately, 50 million people are affected by drought every year and 16 per cent of the country's landmass is drought prone (Fig. 5). In fact, drought is a significant environmental problem too as it is caused by less than average rainfall over a long period of time. In India, about 68 per cent of total sown area of the country is drought prone (Krishnan, 1996; Verma, 1996).

Most of the drought prone areas identified by Government of India lie in the arid, semi-arid and sub-humid areas of the country. The rare droughts of most severe intensity occurred, on an average, once in 32 years and almost every third year is a drought year. During the drought of 2002, Rajasthan faced the worst drought caused by the failure of the South-west monsoon in all the 32 districts. All districts received scanty/ deficient rainfall. Overall monsoon rainfall deficiency in Rajasthan in 2002 was (-) 65 per cent, i.e., 65 per cent below normal, which was the lowest in the last 100 years. In western parts of Rajasthan, the situation was worse with monsoon rain deficiency being 71 per cent below normal.

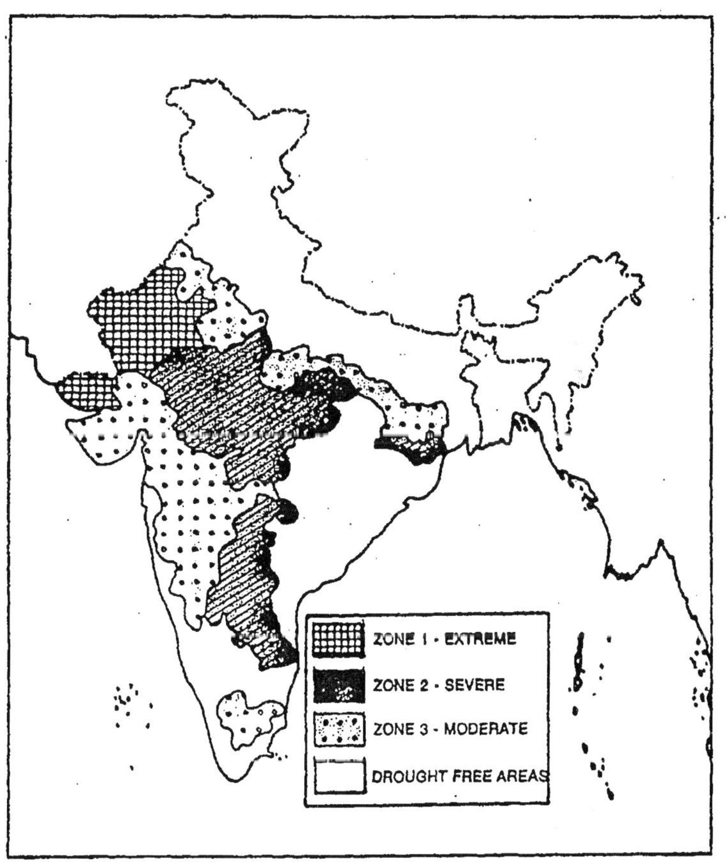


**Fig. 5 : Drought prone area of India**

In 40,990 villages, 44.80 million people and 45.20 million livestock were in the grip of drought of rare severity. Almost entire crop was damaged due to long spell throughout July 2002. As against the normal sown area of 12.90 million hectares in *kharif,* only 8.23 million hectares could be sown. Most of the surface water sources were dry, ground water table was depleted by 3.2 to 6 m and reservoirs received only 32 per cent water as against their designed capacity. Drought 2002 was the fifth year of drought in succession in Rajasthan and a drought of rare severity. Table 4 shows the impact of droughts from 1998 to 2002.

**Table 4 : Impacts of Droughts in Rajasthan**

| *Year* | *No. of Affected Districts* | *No. of Affected villages* | *Population Affected (million)* | *Cattle Affected (Million)* | *Damaged crop (value in Rs. Crore)* |
|---|---|---|---|---|---|
| 1998-1999 | 20 | 20,069 | 21.51 | 29.58 | 2283.49 |
| 1999-2000 | 26 | 23,406 | 26.18 | 34.56 | 3407.02 |
| 2000-2001 | 31 | 30,583 | 33.04 | 39.97 | 3511.77 |
| 2001-2002 | 18 | 7,964 | 6.97 | 6.97 | 1252.27 |
| 2002-2003 | 32 | 40,990 | 44.78 | 45.16 | 4414.00 |

**Source:** Department of Agriculture & Co-operation, Ministry of Agriculture, Govt. of India, New Delhi.

The Unprecedented situation was major challenge for the Government because of the vulnerability of the people. The combination of inflation, unemployment and reduced agricultural and industrial output led to depression of the demand in the economy of the state. At the household level, the major impact of drought was on income and consequently poor people were the worst sufferers due to severe food shortage.

Unlike floods, forewarning of drought is not possible through computer-based studies of numerous climatic and meteorological parameters. Only an idea about the nature and pattern of precipitation is accrued for the ensuing year. The practice prevalent in most of the country to combat droughts is to provide relief measures to drought affected people. Other long-term measures include afforestation to enhance the content of air moisture, water harvesting for the replenishment of ground water and rise of water table, evolving new varieties of crops resistant to water stress conditions and practicing dry-land agriculture, onslaught of desertification, watershed development, development of horticulture, silvi-pasture and agro-forestry and revitalization of Drought Prone Area Programs (DPAP), construction of reservoirs, etc. Food-for-Work Programmes would help the poor in drought - affected areas and also give farmers elsewhere some protection against a price crash later (Kulshrestha, 1997).

Local communities have devised indigenous coping mechanism and drought resistant farming practices in many pockets of the country. With renewed emphasis on watershed management the impacts of droughts and floods may be minimized with people's participation. The most notable among these are the National Watershed Development Programme and the Integrated Watershed Development Programme (IWDP) [Mahlawat and Singh, 2003; Singh and Mahlawat, 2001; Singh *et al.*, 2007].

**4. Cyclones/ Storms**

Storm is a generic term used to describe a large variety of atmospheric disturbances ranging from ordinary rain showers and snowstorms to thunderstorms, wind and wind related disturbances such as gales, tornadoes, tropical cyclones and sandstorms. In meteorology, a tropical storm is restricted to a cyclone with a strong low-pressure centre, strong winds accompanied by heavy precipitation and at times; lightening and thunder (Definition taken from Encyclopedia Britannica).

The World Meteorological Organization uses the term 'tropical cyclones' to define weather disturbances of tropical oceanic origin in which winds exceed 63 kmh$^{-1}$. Tropical cyclones/storms are preceded by tropical disturbances or tropical depressions having lower wind speeds. When wind speed exceeds 119 km hr$^{-1}$, the storm reaches hurricane strengths (Box I). Tropical storms with very strong wind have different names depending on where they occur. In the Atlantic and Eastern Pacific, they are known as hurricanes, in the Western Pacific, including the Philippines, as typhoons, in areas near Australia, as Willy-Willy and in the Indian Ocean cyclones.

Storm-related disasters have been increasing in frequency and intensity during the past decade. The high winds that accompany

**BOX 1 : PHASES OF TROPICAL CYCLONES**

| | | |
|---|---|---|
| Tropical disturbance | : | A weather system which gives rise to a specific area of cloudiness and embedded showers and thunderstorms. |
| Tropical depression | : | Definite counter-clockwise wind circulation with maximum sustained winds of less than 63 km $hr^{-1}$. |
| Tropical storm | : | A tropical cyclone system with maximum sustained surface winds greater than 63 km $hr^{-1}$ but less than 119 km $hr^{-1}$ |
| Hurricane | : | A tropical cyclone with wind speed greater than 119 km $hr^{-1}$. |

**Source** : SDRN contribution to COAG Paper, October , 2000.

tropical storms and the resulting floods have a particularly devastating impact on agriculture. Although, on an average, the intensities of hurricanes have hardly changed during the past three decades, their frequency appears to be on the increase. Further, the devastation caused by tropical storms rose enormously during the 1990s, due in part to the increase of population in storm - prone areas.

The most recent World Disaster Report of the International Federation of the Red Cross shows that during 1990-1999 wind storms and flood-related disasters together accounted for 60 per cent of the total economic loss caused by natural disasters (IFRC, 2000). A significant percentage of disaster causalities, in terms of deaths, injuries and people displaced from their homes and livelihoods, is also attributable to storms and floods.

Despite improved flood awareness and cyclone warning measures, the last 10 years have seen a 300 per cent rise in the number of individuals affected by floods and storms, between 1973 and 1997, hurricanes, cyclones, typhoons, storms and tornadoes claimed, on an average, each year an estimated 11,000 lives and made more than 1.1 million people homeless (IFRC, 2000). In Bangladesh alone three storms, four floods, one tsunami and two cyclones killed more than 400,000 people and affected another 42 million during this period (World Bank, 2000).

The World Watch Institute estimated that during the first 11 months of 1998, weather-related disasters caused more than $ 89 billion in economic losses (compared to $ 55 billion during the 1980s), resulted in 32,000 deaths and displaced 300 million people from their homes and livelihood system (http:www.idndr.org). Most of the disasters in 1998, including record floods in China and Bangladesh and Hurricane Mitch in Central America, were attributed to the *El Nino* or *La Nino* phenomena. Although no aggregate quantitative estimates are available, the economic cost of storm-related disasters in 1999 and 2000 was also considerable. For example, the direct and indirect economic cost of the floods in Mozambique caused by Tropical Storms Elyne and Gloria in February and March 2000 is estimated at US $ 1 billion, compared with the country's export earning of only US $ 300 million in 1999. Geographic location, size and weak economics render Small Island Developing States (SIDs) particularly susceptible to the calamity of weather because when damage occurs, it is quickly transformed to a national scale. The adverse impact of storm -related disaster on economic activities and on land and natural resources is felt more so than in other countries.

India has a very long coastline of more than 8000 km, which is exposed to tropical cyclones originating in the Bay of Bengal and Arabian Sea. On an average about 5-6 tropical cyclones originate in the Bay of Bengal and Arabian Sea every year, out of which 2 or 3 may be severe. The east coastline of the country is more prone than the west to cyclones as about 80 per cent of total cyclones generated in the region hit here. There are two definite seasons of tropical cyclones in the North Indian Ocean. One is from May to June and other is from mid-September to mid-December. Cyclones, hurricanes and typhoons, although named differently, describe the same disaster type. The hazard is referred to as cyclone in the Indian Ocean and South Pacific, hurricane in the Western Atlantic and Eastern Pacific and typhoon in the Western Pacific. These disaster types refer to a large-scale closed circulation system in the atmosphere which combines low pressure and strong winds that rotate counter clockwise in the northern hemisphere (Anjaneyulu, 2004).

The tropical cyclones coming from over the Bay of Bengal become hazardous to the east coastal lands of India, viz., West Bengal, Orissa, Andhra Pradesh and Tamil Nadu. The deadliest hazardous cyclone struck the east coast in 1737 and claimed the lives of 300,000 people. Other disastrous cyclones occurred in 1977 (55,000 deaths), 1864 (50,000 deaths), 1839 (20,000 deaths). The November, 1977 cyclone storm with a speed of 175 km per hour struck Andhra coast and generated three massive, 'storm surges'. The biggest surge of 6 m height rushed into the coastal low lying areas upto 20 km inland and, thus, killed 55,000 inhabitants through drowning caused by sudden inundation, made 2,000,000 people homeless, ruined, 1,200,000 hectares of agricultural crops and made most of the coastal land barren and wasteland because of deposition of thick layer of salt on the soils by storm surges. The saline land could by reclaimed only after three years.

The strongest and most notorious cyclone hit the Andhra coast on May 9, 1990. It was 25 times stronger and more disastrous than the deadliest cyclone of November, 1977 but could claim the lives of 598 people only. Besides killing 598 people, it adversely affected 3,000,000 people, rendered 300,000 people homeless, perished 90,000 cattle and caused loss of 1000 crore rupees worth of property. Very low figure of human causalities (598 deaths) in spite of 25 times more intensity of May 1990 cyclone than the latter was particularly because of the advance monitoring and prediction of the cyclone from the time of its formation in the Bay of Bengal off the southern coast of Tamil Nadu on May 5, 1990 (Bose and Saxena, 2000).

The 29th October, 1999 proved a black and killer day for the inhabitants of the coastal region of Orissa when the super cyclone struck the Orissa coast and caused a havoc of mass destruction through its notorious acts from October 29 to 30, 1999. Nearly one-third of Orissa plunged into gloom and despair. Prior to the final assault by this killer cyclone, a strong cyclone already knocked at the door of Orissa on October 17 to 18, 1999 with a velocity of 200 km $hr^{-1}$. This cyclone claimed the lives of 200 people, damaged 460 villages and adversely affected 5,00,000 people in Ganjam district. Some recent cyclones hitting Indian coast are shown in Table 5.

**Table 5 : Recent major cyclones hitting the Indian coast**

| *Event* | | *Characteristics* | | | *Damages* | |
|---|---|---|---|---|---|---|
| *State* | *Time* | *Wind Velocity (km $hr^{-1}$)* | *Sea Surge (m)* | *Lives Lost (No.)* | *Houses damaged (million)* | *People Affected (million)* |
| Andhra Pradesh | May, 1990 | 240-250 | 5-6 | 928 | 1.38 | 7.7 |
| Andhra Pradesh | November, 1996 | 140-150 | 2.2 | 1057 | 0.65 | 7.0* |
| Gujarat | June, 1998 | 200 | 2-3 | 1250 | 0.26 | - |
| Orissa | October, 1999 | 270-300 | 6 - 7 | 10086 | 0.22 | 37.5 |

* Families

**Source** : Sinha, 2001

**5. Landslides**

Landslides are being increasingly viewed as natural hazards. The concern regarding landslide problem can be understood in the light of the fact that a majority of landslides are triggered by natural forces including substantial rainfalls, cloud bursts, volcano eruption, earthquakes, etc., and as such are difficult to predict. Landslide problem has increased in magnitude immensely due to manmade activities as well. On the other hand, floods and droughts do not happen by accident, but are the products of patterns of wind, temperature and precipitation that produce meteorological extremes.

India provides striking examples of a bewildering variety of landslides and other mass movements. There have been numerous landslides events, unique and unparalleled. In the recent memory, the Darjeeling floods of 1968 destroyed vast areas of Sikkim and West Bengal by unleashing numerous landslides, killing thousands of people. These landslides occurred over a three-day period with precipitation ranging from 500 to 1000 mm in an event of a 100-year rainfall return period. The 60 km mountain highway to Darjeeling got cut off at 92 places resulting in total disruption of the communication system. Yet another landslide tragedy of unprecedented dimension was the Alaknanda Tragedy of July 1970 that resulted from the massive floods in river Alaknanda upon breach of a landslide dam at its confluence with river Patal Ganga. More recently, the Malpa rock avalanche tragedy, hit headlines as it instantly killed 220 people and wiped out the entire village of Malpa on the right bank of river Kali in the Kumaun Himalaya. Landslides in the Southern India also received public imagination when the recent Amboori landslide in the State of Kerala killed 23 people. Sinha (2006) reported that landslides do occur in a loose cover of debris consisting of boulders. The major landslides in the Nilgiri hills are the Runnymede landslide, the Glenmore slide, the Conoor slide and the Karadipallam slide. The first small scale landslide hazard map of India (1 : 6 million scale) was-published by Building Materials Technology Promotion Council of the Government of India.

Other disasters like forest fires, insect - pests infestation and outbreak of diseases, train accidents and plane crash take place but the severity is not critical. The peculiarity of disaster scenario of India is multi-hazard vulnerability of number of states. The States of Gujarat, West Bengal and North-east States are highly prone to multiple hazards, e.g., the Kachchh district of Gujarat was going under severe drought condition when a major earthquake hit that area and the same area faced sever cyclone in 1998 as well.

## Vulnerability

Vulnerability to disasters is a function of human action and behaviour. It describes the degree to which a socio-economic system or physical assets are either susceptible or resilient to the impact of natural hazards. It is expressed on a scale from 0 (no damage) to 1 (total loss). It is determined by a combination of several factors, including awareness of hazards, the condition of human settlements and infrastructure, public policy of administration, the wealth of a given society and organized abilities in all fields of disaster and risk management.

Lack of awareness among, the public decision-makers about factors and human activities that contribute to environmental degradation and disaster vulnerability are aggravating these trends (Box II).

| **Box II : Factors and human activities that contribute to environmental degradation** | | |
|---|---|---|
| **Causes and Limits** | **Economic Development** | **Social Development** |
| Disaster limits developments | Destruction of fixed assets. Loss of production capacity, market access or material inputs. Damage to transport, communications or energy infrastructure. Erosion of livelihoods, saving and physical capital. | Destruction of health or education infrastructure and personnel, Death, disablement or migration of kcy social actors lcading to an erosion of social capital. |
| Development Causes disaster risk | Unsuitable development practices that create wealth for same at the expense of unsafe working or living conditions for others or degrade the environment. | Development paths generating cultural norms that promote social isolation or political exclusion. |
| Development reduces disaster risk | Access to adequate drinking water, food, waste management and a source dwelling increases peoples resilience. Trade and technology can reduce poverty. Investing in financial mechanisms and social security can cushion against vulnerability. | Building community cohesion recognizing excluded individuals or social groups (such as women) and providing opportunities for greater involvement in decision making enhanced educational and health capacity increases resilience. |

The fast pace of growth and expansion without comprehensive understanding of safety nets has brought forth a range of issues that can put man and his surroundings at a risk of prospective hazards. The safeguards of existing systems are limited and the risks involved are high. For these, a sound understanding of the vulnerability concept and risk factor is essential (Sinha, 1998).

The concept of vulnerability therefore, implies a measure of risk combined with the level of social and economic ability to cope with the resulting event in order to resist major disruption or loss (Fig. 5). The susceptibility and vulnerability to each type of threat will depend or their respective differing characteristics (Kumar and Parmar, 2005).

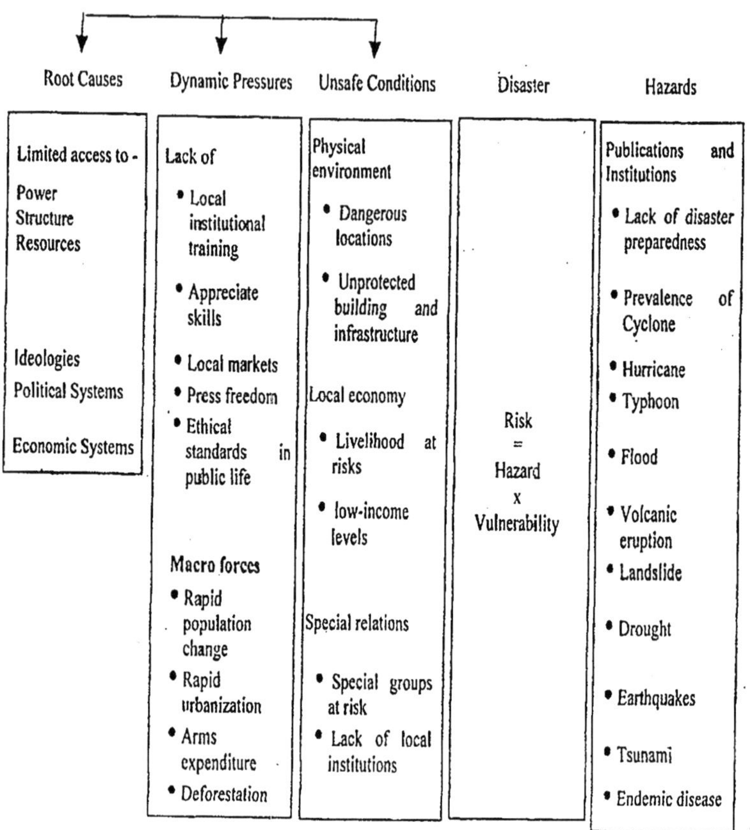


**Fig. 5: The model progression of vulnerability with three main levels, root causes, dynamic pressure and unsafe conditions**
**Source** : Blakie *et al.*, 2004

Due to a large number of most vulnerable regions in India, natural disasters management has emerged as a high priority for the country. Going beyond the historical focus on relief and rehabilitation after the event, we now have to look ahead and plan for disaster preparedness and mitigation in order that the periodic shocks to our development efforts are minimized (GOI, 2004).

**Risk:** Definitions of risk in the hazards literature vary from those that equate risk with probability to those that see risk as the product of a probability and a particular kind of impact occurring. According to ADPC (2002), risk is the likelihood or probability of a hazard event of a certain magnitude occurring. Risks are measures of the threat of hazards.

Understanding the interactions of physical and social processes is fundamental to estimating future risk and the value of prevention. The following equation is commonly used to effect the component of risk :

***Risk*** = Hazard x vulnerability x Amount

Where,

***Hazard*** = Probability of event with certain magnitude

***Vulnerability*** = Degree of damage which is function of magnitude of event and type of elements at risk

***Amount*** = Quantification of the elements at risk, e.g., replacement costs of buildings, infrastructure etc.; loss of function or economic activities, and number of people.

The figure 6 illustrates essentially the four factors, viz., hazards, location, exposure, and vulnerability, which contribute to risk. They are :

**Fig. 6 : Factors that contribute to risk**

With climatic variability, deteriorating environment, burgeoning population and increased frequency of sudden catastrophes, a large proportion of humanity is constantly at risk. We have seen that the super cyclone in Orissa, earthquake in Gujarat and Tsunami in the Andaman & Nicobar Islands and the eastern coast of the country have devastating effects and has also negated years of efforts invested towards development. Mitigating disaster risks and making the nation and the communities within it resilient is the requirement one needs to work towards. Vulnerability needs to be analyzed in the face of hazards. The interaction between vulnerability and exposure to hazards define risk. To identify and know the risk is the first step towards mitigation. Effective risk reduction involves mitigation measures in hazard prone areas. It may also involve overcoming the socio-economic, institutional and political barriers to the adoption of effective risk reduction strategies and measures in developing countries. The systematic development and application of policies, strategies and practices to minimise vulnerabilities, hazards and the unfolding of disaster impacts throughout a society, in the broad context of sustainable development. Risk reduction may take place in two ways :

### (a) Preparedness

Indeed, it has been noticed in the past that as and when attention has been paid to adequate preparedness measures, the loss to life and property has considerably reduced. The pinnacle of disaster preparedness is a disaster resilient community. In most communities emergency management involves an authoritative and systematic approach to co-ordinate all phases of a disaster. This protective process embraces measures that enable governments, communities and individuals to respond rapidly to disaster situations to cope up with them effectively. Preparedness includes the formulation of viable emergency plans, the development of warning systems, the maintenance of inventories and the training of personnel. It may also embrace search and rescue measures as well as evacuation plan for areas that may be at risk from a recurring disaster (GOI, 2005).

Preparedness, therefore, encompasses those measures taken before a disaster event which are aimed at minimizing loss of life, disruption of critical services, and damage when the disaster occurs. All preparedness planning needs to be supported by appropriate legislation with clear allocation of responsibilities and budgetary provisions (Sahni and Madhavi, 2003).

### (b) Mitigation

A quick rescue and relief mission is an essential necessity when disaster strikes. In addition to effective response system, adequate preparedness levels can minimise damage to a considerable extent. According to the Indian Act 2005, "mitigation means measures aimed at reducing the risk, impact or effects of a disaster or a threatening disaster situation". There is the urgent needs to ensure that disaster mitigation strategies get enmeshed and integrated with every development process. Mitigation embraces all measures taken to reduce both the effect of the hazards itself and the vulnerable conditions to it in order to reduce the scale of future disaster. Therefore, mitigation activities can be focused on the hazard itself or the elements exposed to the threat. Examples of mitigation measures which are hazard specific include modifying the occurrence of the hazard, e.g., water management in drought prone areas, avoiding the hazard by sitting people away from the hazard and by strengthening structures to reduce damage when a hazard occurs. In addition to these physical measures mitigation should also be aimed at reducing the physical, economic and social vulnerability to threats and the underlying causes for this vulnerability.

In the recent years, the disaster management set up in India has oriented itself towards a strong focus on preventive approaches, mainly through administrative reforms and participatory methods.

**Disaster Management**

The World today is living with more risks and uncertainties than ever before. With climatic variability, deteriorating environment, burgeoning population and increased frequency of sudden catastrophes, a large proportion of humanity is constantly at risk. Recent years have witnessed an alarming increase in the frequency of disaster occurrence as well as the magnitudes of their impacts. We have seen that the super cyclone in Orissa, earthquake in Gujarat and Tsunami in the Andaman & Nicobar Islands and the eastern coast of the country have had devastating effects and have also negated years of efforts invested towards development. Mitigating disaster risks and making the nation and the communities within it resilient is the requirement one needs to work towards. Vulnerability needs to be analyzed in the face of hazards. The interaction between vulnerability and exposure to hazards define risk.

Disaster management can be defined as the body of policy and administrative decisions and operational activities that pertain to the various stages of a disaster at all levels. Disaster management comprises of two components, viz., Disaster preparedness and Disaster Relief. Both the components have their own importance. In the aftermath of a disaster, relief and rehabilitation need immediate attention and early response. A relatively smaller investment in disaster preparedness can save thousands of lives and vital economic assets, as well as reduce the cost of overall relief assistance. The disaster preparedness is a very cost-effective component of disaster management (Sinha, 1998; Carter, 1991).

According to the Indian National Disaster Management Act 2005, disaster management means a continuous and integral process of planning, organizing, coordinating and implementing measures which are necessary or expedient for (i) prevention of danger or threat of any disaster (ii) mitigation or reduction of risk of any disaster or its severity or consequences (iii) capacity building (iv) preparedness to deal with any disaster (v) prompt response to any threatening disaster situation or disaster (vi) assessing severity or magnitude of effects of any disaster (vii) evacuation, rescue and relief, and (viii) rehabilitation and reconstruction.

The High Powered Committee defined Disaster Management as "a collective term encompassing all aspects of planning for and responding to disasters, including both pre - and post-disaster activities. It may refer to the management of both the risks and consequences of disasters."

Clearly, the term management has emerged as an umbrella term that encompasses the entire disaster cycle, including mitigation. This needs careful noting and widespread awareness because traditionally the term management was restrictively used to address only post disaster.

Disaster management in India, thus, was much focused on relief, rescue, rehabilitation and recovery. There is now a shift to a new disaster management paradigm that stresses on prevention, mitigation and preparedness, while strengthening its emergency response (rescue, relief, rehabilitation and recovery). The new cyclic paradigm has six phases covering the pre-disaster activities. The former includes prevention, preparedness and mitigation, while the latter consists of emergency response (rescue and relief), rehabilitation and recovery (reconstruction).

According to IDNDR Report on Technology for Disaster Reduction undertaken by IDNDR Programme Forum 1999 operations stated that "unless the old mindsets get changed, the cause of disaster mitigation will continue to suffer at the hands of traditional disaster managers".

Broadly disaster management can be divided into pre-disaster, during and post-disaster stage. For better management, the disaster management cycle for a vulnerable community has to be alert and effective in all the three stages as follows :

(a) **Pre-disaster Stage** : Activities are taken to reduce losses caused to human lives and property by the hazard and ensure that these losses are also minimized when the disaster strikes. Risk reduction activities are taken under this stage and they are termed as mitigation and preparedness activities.

(b) **During Disaster Occurrence Stage** : Activities taken to ensure that needs and provisions of victims are met and suffering is minimized.

Activities taken under this stage are called as emergency response activities.

(c) **Post-Disaster Stage** : Activities taken to achieve early and durable recovery and does not expose the earlier vulnerable conditions in affected area. The objective is to bring the affected community back on rails as quickly as possible. Activities taken under this stage are called as response and recovery leading to development.

The different stages of disaster management are reflected as a disaster management cycle (Fig. 7). To begin with, a disaster event occurs and creates the associate impact. The duration of the event depends on the type of hazard, for example, during an earthquake, ground tremors may occur for seconds, while flooding may take place over a longer sustained period.

Within each phase of the disaster management cycle, short-range goals can simultaneously contribute to long-range ones, such as strengthening people's capacities to withstand disasters. For example, the reconstruction of water supplies should merge naturally into on-going development activities (such as community mobilization) to further improve the water-supply systems (or other agreed environmental health goals). During "normal" times, these health development activities should aim to reduce the vulnerability of people and infrastructure to future emergencies and disasters. Thus, the routine construction of water works should, for example, incorporate design features that protect them from known hazards. Community participation in each phase strengthens the organizational structure of neighbourhoods and localities, and improves methods of providing early warning of hazards, of planning emergency responses, etc.

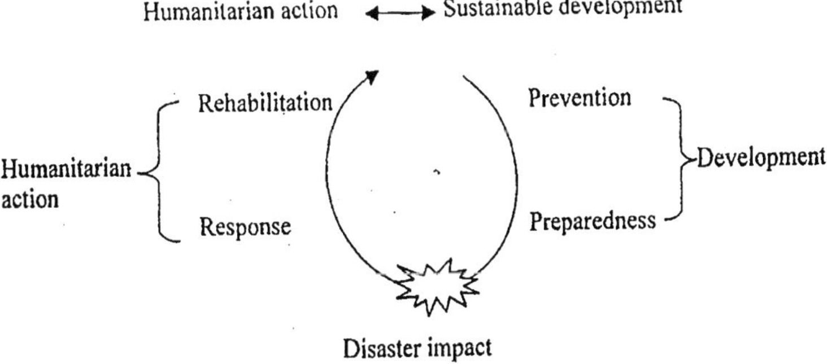


**Fig. 7 : Disaster management cycle**

The concept of integration is central to current thinking and practices of disaster management. There are at least four ways in which the integration of efforts is important:

(i) It is necessary for different sections to work together,

(ii) Environmental health planning must be viewed as part of overall health planning,

(iii) Rural and urban communities must participate in all phases of emergency relief and development, and

(iv) A responsive and accountable system of professionals and volunteers is needed.

Response and relief refer to the first stage response to any disaster that includes setting up control rooms, putting the contingency plan in action, issuing warning, taking action for evacuation, i.e., taking people to safer areas, and rendering medical aid simultaneously providing food, drinking water, clothing, etc. to the homeless, restoration of communication and disbursement of assistance in cash or kind.

The recovery stage includes activities that encompass the three overlapping phases of emergency, relief, rehabilitation and reconstruction. Emergency relief activity such as immediate relief, rescue, damage assessment and debris clearance are undertaken during and immediately following a disaster. Preventive measures should also be taken for the spread of diseases. Rehabilitation includes the provision of camps with temporary public utilities and shelter as interim relief measures for long-term recovery. Reconstruction efforts are made to return the communities to improved pre-disaster functioning that includes repair and reconstruction of buildings, infrastructure and lifeline facilities so that long-term development prospects are enhanced.

The development process stage is an ongoing activity, which has to be resumed. It deals with long-term prevention/disaster reduction measures. In this regard, construction of embankments against flooding, irrigation facilities as drought proofing measures, land use planning, construction of houses capable of withstanding the onslaughts of future disasters are taken up as part of development plans.

Thereafter comes the stage of prevention and mitigation which refers to such preventive measures during disaster free period that would reduce the impact if a disaster reoccurs. Thus, mitigation embraces all measures taken to reduce both the effect of the hazard itself and the vulnerable conditions of the community in order to reduce the adverse impacts of a future disaster. Lastly, the preparedness process embraces measures that enable the government, community and individuals to respond rapidly to disaster situation in order to cope up with them effectively. It includes the formulation of viable emergency plans, development of warning system, maintenance of inventories, mock drills and training of personnel.

For effective disaster management, it is essential to have workable action plans at all levels from community level upwards. The aim is to have a tested and tried system -in place, which comes into action at very short notice. This system becomes a powerful tool for risk reduction.

Thus, disaster management is the continuous process by which all individuals, groups and communities manage hazards in an effort to avoid or ameliorate the impact of disasters resulting from the hazards. Actions taken depend in part on perceptions of risk of those exposed. Effective disaster management relies on integration of emergency plans at all levels of government and non-government involvement. Activities at each level affect the other levels. It is common to place the responsibility for governmental disaster management with the institutions for civil defence or within the convenient structure of the emergency services. In the private sector, disaster management is sometimes referred to as business continuity management.

## GOVERNMENT INITIATIVES TO MANAGE THE DISASTERS

### (a) Nodal Ministries for Managing Disasters

The disaster management structure of the country is such that number of Ministries and Departments of the Government are handling the disaster situation arising due to different types of hazards independently (Table 6). The Department of Agriculture & Co-operation (DAC) Ministry of Agriculture, New Delhi is the Nodal Department for all matters concerning natural disaster at the centre. In the DAC, the Central Relief Commissioner functions as the Nodal Officer to co-ordinate relief operations for all natural disasters. The Central Relief Commissioner functions receives information relating to forecast /warning of the natural calamity from the Director General, India Meteorological Department (IMD) or from the Central Water Commission (CWC) on a continuous basis. Besides, he also monitors the developments taking place and provides the necessary feedback, through the Secretary to the Minister for Agriculture and Prime

Minister. Various other committees working at the national level for the Disaster Management Committee (NCMC) and Crisis Management Group.

**Table 6 : Nodal ministries for managing different types of disaster**

| *Type of Disaster/Crises* | *Nodal Ministry* |
|---|---|
| Air Accidents | Ministry of Civil Aviation |
| Civil Strife | Ministry of Home Affairs |
| Railway Accidents | Ministry of Railways |
| Natural Disasters except drought | Ministry of Home Affairs |
| Drought | Ministry of Agriculture |
| Chemical Disasters | Ministry of Environment |
| Biological Disasters | Ministry of Health & Family Welfare |
| Nuclear Accidents | Department of Atomic Energy |
| Major breakdown of any of the essential services posing problems | Concerned Ministries |

**Source** : Disaster Management in India: A Status Report - 2004, National Disaster Management Division, Ministry of Home Affairs, Govt. of India, New Delhi.

To cope up with and manage the situation arising out of frequently occurring natural disasters a coordinated disaster mechanism exists within the government framework. The essential responsibility of disaster management lies with the State Government where the disaster has occurred. In the event of disaster which are spread over several States and with uncontrollable proportions, the Central Government may be required to supplement the efforts of State Government by taking appropriate rescue and relief measure. The policy and arrangement for financing the State Governments to provide relief and rehabilitation measures in areas affected by natural calamities are governed by the recommendations of the Central Finance Commissions appointed from time to time. Under the present scheme a Calamity Relief Fund (CRF) has been constituted for each State with contribution from the Central and State Governments to undertake relief and rehabilitation measures. The annual allocations of the CRF to the various States are based on their trend of expenditure on natural calamities during the past 10 years. As per the XI Finance Commission recommendations the Central share of CRF is released to the State Governments in two installments. As recommended by the present Finance Commission a committee of experts and representatives of the states has drawn up a list of items, expenditure on which alone will be chargeable to the CRF. A State Level Committee headed by the Chief Secretary decides the norms of assistance under each of the approved schemes.

Government of India has also created a Contingency Action Plan (CAP) for dealing with contingencies arising in the wake of natural disasters, which has been periodically upgraded. The major emphasis in the Plan is on providing relief in response to the occurrence of a major calamity due to any of the natural hazards which could not be dealt with by the states through their own resources. The National Contingency Action Plan facilitates launching of relief and rescue operations without any delay. The CAP identifies initiatives required to be taken by various Central Ministries and Public Departments in the wake of natural calamities and sets down the procedures and determines the focal points in the administrative machinery. According to the CAP, primary relief function of the Central Government includes the following :

(i) Forecasting and operation of warning system,

(ii) Maintenance of uninterrupted communication,

(iii) Wide publicity to warning of impending calamity, disaster preparedness and relief measures through TV, radio and newspapers,

(iv) Transport with particular reference to evacuation and movement of essential commodities and supply of petroleum products,

(v) Ensuring availability of medicines, vaccines and drugs,

(vi) Preservation and restoration of physical communication links; and

(vii) Mobilization of financial resources.

In the federal set-up of India, the responsibility to formulate the Governments to a natural calamity is essentially that of the concerned State Government. However, the Central Government, with its physical and financial resources does provide the needed help and assistance. The dimensions of the response at the level of Central Government are determined in accordance with the existing policy of financing the relief expenditure and keeping in view the factors like : (i) The gravity of a natural calamity, (ii) The scale of the necessary relief operation, and (iii) The requirements of central assistance for augmenting the financial resources at the disposal of the State Government.

There has been improvement in the government response to natural disasters in terms of its effectiveness. This is chiefly due to the emergence of well organised administrative machinery, presence of relief manuals at district level, predetermined allocation of duties and recognized public private partnerships. However, the absence of the integrated policy at national level has led to overlooking of some of the vital aspects of disaster management (Sinha, 2001).

**(2) International Efforts in Management of Disasters**

It is a truism that the largest number of affected people due to disasters live in Asia and Africa, the two poorest continents. The threat posed by the natural disasters on the economies of the nations, prompted the United Nations General Assembly through its resolution 44/236, 1989 to proclaim the last decade of the last century as the International Decade for Natural Disaster Reduction (IDNDR) from 1990-1999. The concept of IDNDR was based on the fact that proper use of the available knowledge and expertise to mitigate risks arising due to various kinds of natural disasters be made.

The purpose of the decade was to marshal the political resolve, experience and expertise of each country to reduce the loss of life, human sufferings, and economic losses from natural disasters. According to IDNDR targets, by the year 2000, all nation would have achieved :

**(a) Complete National Risk Assessments**

(i) Identify the natural hazards, which pose the threat of disaster,

(ii) Evaluate the geographic distribution of hazard threat and their frequency and impacts,

(iii) Assess the vulnerability of the most important concentrations of the populations, development and resources

**(b) Implementation of National and/or Local Prevention and Preparedness Plans**

(i) Adopt land use and construction practices that will reduce risks,

(ii) Adopt emergency - response plans that identify 'hazard scenarios', essential action and responsible organizations,

(iii) Implement awareness programmes to educate people about risks and training programmes to improve the capabilities of the responsible persons, and

(iv) Implement concrete measures to mitigate damages and strengthen people's resilience to hazards.

**(c) Implement global, Regional, National and Local Warning Systems**

(i) Implement systems for monitoring and predicting threatening phenomena in time to reduce impacts,

(ii) Implement communication system for disseminating warnings, and

(iii) Educate decision - makers about hazards and option for reducing disaster risk.

The Yokohama Strategy, adopted in May 1994, called upon all countries to unequivocally give political commitment to reduce their vulnerability through appropriate means. The strategy contributed to shifting focus from disaster response to prevention and mitigation.

Realizing the need to continue the momentum developed during the IDNDR at national level, the Government of India is observing the current decade as the "National Decade for Disaster Reduction" (NDDR). After the completion of IDNDR, Government of India decided to observe October, 29 as NDDR (National Day for Disaster Reduction) every year to commemorate the day a Supercyclone had hit Orissa in 1999. The basic objective of NDDR is to create public awareness and educate the people about the impact of natural disasters and importance of disaster preparedness and mitigation (Satendra and Gupta, 2003).

In the light of above message and to meet the objectives stated in the IDNDR, the Government of India started number of important programmes. Various initiatives taken at national, state and local levels are

**1. High Power Committee on Disaster Management Plans**

A High Power Committee (HPC) was set up at the initiative of the Prime Minister to look into the issue of Disaster Management planning at National, State and District levels under the Chairmanship of Mr. J.C. Pant, Former Secretary, Government of India. The term of reference of HPC include : (i) review of existing arrangements for preparedness and mitigation of natural and man-made disasters, (ii) recommendation for strengthening existing organisational structure; and (iii) preparation of model disaster management plans at National, State, and District levels. Five sub-groups were constituted under the HPC to look into the various aspects of disaster management planning, e.g., water and climate related hazards, geological hazards, chemical/industrial/nuclear disaster; accident related disasters, biological disasters, etc.

The main emphasis laid by HPC is on mitigation and preparedness of disasters. The committee has had a number of consultations with various groups, including academicians, technocrats and voluntary agencies to arrive at a common plan of action. The role played by the NGOs in the post disaster management is evident to every one. Clear identification of the role and responsibilities of NGOs has already been completed. To use the traditional wisdom and its integration with scientific knowledge for disaster mitigation, the HPC has organized meetings for developing a network of knowledge based organizations. The Committee has already submitted its report in 2001 (Sinha, 2001).

### 2. National Committee on Disaster Management

The Government of India has established a National Committee on Disaster Management under the Chairmanship of the Prime Minister. The Committee has the representation of major National and State political parties. This committee is supposed to suggest the institutional and legislative measures for strengthening the existing disaster management structure of the country.

### 3. National Centre for Disaster Management (NCDM)

The Institute of Public Administration (IIPA), New Delhi recognized the need for systematic disaster mitigation efforts at national level in 1993. In response to a proposal submitted by the IIPA to the Ministry of Agriculture, Government of India, National Centre for Disaster Management was established at the IIPA in 1995. The NCDM functions as a Nodal Centre in the country for human resource development in the area of disaster management. The main objectives of the Centre are : (i) to prepare an exhaustive information base on damage caused and resources spent on mitigation practices and relief work for various types of natural disasters ; (ii) to establish links between the Nodal Ministry of natural disasters, technical assistance and services to the national programmes on natural disaster management; and (iii) to assist South-east Asian -and SAARC countries in formulating their policies and developing their capabilities in all aspects of disaster management as and when assigned by the Nodal Ministry. The Centre is doing networking with various Central/State Government Ministries, training Institutions, Autonomous Organizations and Universities. The NCDM collaborates and assists these organizations in organizing training programmes, workshops, seminars and research activities, etc. related to disaster mitigation and management.

**4. United Nations Initiatives for Vulnerability Reduction and Sustainable Environmental Management**

In the area of disaster Management, the United Nations General Assembly has given United Nations Development Programmes (UNDP) the mandate of being United Nations focal point to support and strengthen national capacities related to disaster mitigation, prevention and preparedness. The interventions in disaster mitigation within the framework of country co-operation programmes are aimed at ensuring a sustainable recovery in the aftermath of a disaster as well as reducing people's vulnerability to natural disaster .

In India, UNDP is the convener for the United Nations Disaster Management Team (UNDMT), which is an inter-agency working group comprising of FAO, ILO, UNDP, UNESCO, UNFPA, UNICEF, UNIDO, UNIFEM, UNV, WFP and WHO. UNDMT acts as the coordinator for disaster - related response activities within the UN system and works in collaboration with the National Government to support disaster management activities in the country. The global objectives of UNDP in vulnerability reduction and sustainable environment are : (i) to achieve a substantial reduction in disaster risks and protection of development gains, (ii) to reduce the loss of life and livelihoods due to disasters; and (iii) to ensure that disaster recovery serves to consolidate sustainable human development. UNDP has been supporting various initiatives of the Central and State Governments in India to strengthen the disaster management capacities.

Vulnerability Reduction and Sustainable Environmental Management are one of the four thematic areas for Government of India (GOI) - UNDP Country Co-operation Framework (2003-2007), with two broad projects : One focusing on sudden disasters and another one on slow brewing up disasters. These projects strengthen the decentralization processes like strengthening of States' capacities and local self governments and promote the gender equality, which are the two objectives of United Nations Development Assistance Framework, in vulnerability reduction and sustainable environmental management. A community based approach to achieve the objectives set forth by the Central Government and would strive to integrate disaster management in the national and state development agenda as well as enhance the capacity of government functionaries to achieve the objectives. These two national programmes, provide a platform for promoting partnerships with national and international counterparts for developing co-operation, to support Government -led initiatives for disaster mitigation and vulnerability reduction.

**5. National Capacity Building Project in Disaster Management**

In 1996, UNDP launched the National Capacity Building Project to enhance the capabilities of the Government for disaster management. The project aimed at strengthening the National Centre for Disaster Management (NCDM), training government functionaries and faculty of Disaster Management cells in the State Institutes of Public Administration. The strategy was to develop NCDM as a resource centre to produce manuals and guidelines; formulate policy and disaster management strategy proposals. The project has helped the government to have a core group of government officers trained in disaster management, capable of formulating policies and developing a National Disaster Management Plan. These officers acted as resource persons for the High Powered Committee, established by the Prime Minister of India to revamp the disaster management structure in the country. The control room at the Natural Disaster Management Division located within the Union Ministry of Agriculture was strengthened with essential equipments that would help the Government of India to better prepare for disasters through early warning and assessment missions during disasters (UNDP, 2004).

UNDP - supported initiatives in the aftermath of the two major disasters - the 1999 Supercyclone in Orissa and the 2001 devastating earthquakes in Gujarat demonstrated the effectiveness of development of approaches to disaster risk mitigation and vulnerability reduction in India. Integration of community-based disaster preparedness and mitigation plans into the development plans prepared by local governments, and strengthening of local capacities and institutions, have proved to be successful in mitigating the impact of subsequent disasters.

**6. Strengthening of Information and Communication Technology in Disaster Management**

Disaster whether natural or man-made play havoc with the lives of millions of people every year. Their aftermath is nothing but a grim picture of death, destruction and sufferings. River flood, most devastating disaster in India, is mainly caused due to the variations in rainfall, climate and topography. It is not always possible to avoid disasters but the sufferings can be minimized by proper disaster management through appropriate disaster management tools. One of the disaster management tools is Information and Communication Technology (ICT). The advancement in ICT is in the form of Internet, GIS, Remote Sensing, Space Technology, etc. can help a great deal in planning and implementation of disaster reduction measures as well as warning, relief, rescue, rehabilitation and mitigation.

**Prevention and Mitigation : Future Perspectives**

An early rescue and relief is prerequisite when disaster strikes. In addition to effective response system, adequate preparedness measures can minimize damage to a considerable extent. Indeed, it has been noticed in the past that as and when attention has been paid to adequate preparedness measures, the loss to life and property has considerably reduced. Going along this trend, the disaster management set up in India has, in recent years, oriented itself towards a strong focus on preventive approaches, mainly through administrative reforms and participatory methods. Preparedness measures such as training of role players including the community, development of advanced forecasting systems, effective communications, and above all a sound and well networked institutional structure involving the government organizations have greatly contributed to the overall disaster management in the country.

The approach of reducing community vulnerability for reducing disasters has paid right dividends. The first step in this direction has been of identification of vulnerable communities. The communities which are periodically exposed to natural hazards, and with low levels of coping capacity especially the poor, need immediate attention. Marginal sections of rural communities and dwellers of informal settlements and slums in urban areas are the worst suffers.

Efforts in the direction of integrating disaster prevention into habitat planning processes are one of the most viable disaster prevention means. The National Centre for Disaster Management's work on developing and testing methods or integrating risk reduction using community participation into urban planning is one such initiative. The general direction of current efforts is one of multi-pronged approach towards mobilization of community perceptions and a culture of prevention of natural disasters (Sinha, 2001).

In the era of information and space technology disasters present an opportunity to learn and innovate at national and international level. The need to learn from frequently occurring natural disasters can be emphasized with the help of examples of Andhra Pradesh cyclone, 1977 and Maharashtra (Latur) earthquake, 1993 which led to the preparation of state - specific disaster management plans without much learning at national level. As a consequence the nation had to pay a very heavy price during Orissa cyclone (1999) and Gujarat earthquake (2001). Similarly, the best practices need to be replicated at International level where this learning process from one disaster can be enhanced to minimize the losses elsewhere in the world. Technological innovations can be used very effectively for this purpose (Ajai and Kimothi, 2003; Chakraborti *et al.*, 2003; Ganesh Raj and Nijagunappa, 2003, 2004; Gaur, 2003; NNRMS, 2003; Singh *et al.*, 2003; Sinha, 2003).

## CONCLUSION

Disasters have always affected mankind and will probably continue as long as life exists on this planet. When occur, they leave a long trail of sufferings. Multiple deaths, destruction, injuries, losses cause trauma among the survivors. This exists all over the world and India is one of the most vulnerable regions. Floods, droughts, cyclones, earthquakes and landslides have been recurrent phenomena in our country. Thousands of lives are lost and properties worth crores are damaged. About 60 per cent of the landmass is prone to earthquakes of various intensities. Over 40 million hectares, is prone to floods; about 8 per cent of the total area is prone to cyclones and 68 per cent of the area is susceptible to drought. In addition, changes in the global environment threaten us with the possibility of severe cyclones, rising sea levels, droughts, etc. Today's society is becoming ever more rapidly vulnerable to natural disasters due to the concentration of population in mega cities.

Although, it is almost impossible to fully recoup the damage caused by the disasters, yet it is possible to minimize the potential risk by developing early disaster warning strategies, preparing and implementing development plans to provide resilience to such disasters, and helping in rehabilitation and post disaster reduction. Disaster response and recovery efforts require timely interaction and coordination of public emergency services in order to save lives and property.

Disaster management is the continuous process by which all individuals, groups and communities manage hazards in an effort to avoid or ameliorate the impact of disasters resulting from the hazards. Effective disaster management relies on thorough integration of emergency plans at all levels of government and non-government involvement.

There has been a paradigm shift in our approach to disaster management during the last one decade. This shift is from a relief centric approach to a multi-dimensional endeavour involving diverse scientific, engineering, financial and social processes to adopt a multi-disciplinary and multi-sectoral approach with stress on building up capabilities of community to enable them to work towards their own risk reduction. The role of NGOs in this context therefore, assumes added significance.

Disasters both natural and manmade negatively impact sustainable development. The vulnerability to disasters should have the resources, capacities and opportunities to cope and recover from stresses and shocks. The need of the hour is therefore, the integration of scientific, technological, ecological, economic and social dimensions of disaster management with sustainable management.

In addition to a huge toll of human lives and property, natural disasters severely hamper development. It may not be possible to prevent natural disasters totally, but with careful pre-disaster planning and preparedness, we can reduce the adverse effect of the hazards and the economic loss. Here evolves the significance of disaster management. Disaster management means a planned and systematic approach towards understanding and solving problems in the wake of disasters. It involves the systematic observation and analysis of measures relating to disaster prevention, mitigation, preparedness, emergency response, rehabilitation and reconstruction. Disaster prevention and preparedness is a neglected aspect even though the occurrence of disaster is a regular phenomenon. Risk and resilience may be vague concepts, but when rooted or institutionalized within the community based organizations risk reduction becomes reality. It may be realized that disaster management is a function of community preparedness. A natural hazard can transform into a natural disaster depending on its impact on society in terms of life and property. In order to enhance community preparedness, a proper safety plan is essential. The community preparedness plan involves all pre-disaster planning to reduce the losses as a result of natural disaster. It is basically a synthesis of various specific plans to solve a common purpose.

## REFERENCES

ADPC (2002). *Community Based Disaster Management : Trainer's Guide.* Asian Disaster Preparedness Centre, Bangkok, Thailand.

Ajai and Kimothi, M.M. (2003). Forest fire management. *In : Bulletin of the National Natural Resources Management System,* NNRSM (B)-28 (Eds. V. Jayaraman, R.R. Navalgund, S.K. Bhan, S.K. Shivkumar). Department of Space, Government of India, Bangalore, pp. 95-107.

Akkihal, A.R. (2006). Inventory pre-positioning for humanitarian operations, Edgar E. Blanco, Ph.D., Massachusetts Institutes of Technology, Cambridge, USA.

Alexander, D. (1993). *Natural Disasters.* UCL Press, London.

Anjaneyulu, Y. (2004). TRMM and SMM/I based modeling studies on heavy rain and flooding episodes from the Mahi river basin mainland, Gujarat *J. Indian Geophys. Union,* 8(1): 39-48.

Anonymous (2005). Tsunami disaster - Relief and rescue operations. *Employment News,* 29(43): 56.

Blaiki, P.M. Cannon, T., Davis, I. and Wisner, B. (Eds.) [1994]. *At Risks : Natural Hazards, Peoples Vulnerability and Disaster.* New York, USA.

Bose, S. and Saxena, N. (2000). Tropical Cyclones. *Employment News,* 25(4) : 1-3.

Bruce, A. (1978). *Earthquakes : A Primer.* W.H. Freeman and Company, USA, p. 240.

Carter, N. (1991). *Disaster Management : A Disaster Manager's Handbook.* Asian Department Bank, Manila, Philippines.

Chakraborti, A.K. Bhanumurthy, V. and Chandrashekar, G. (2003). Remote Sensing towards filling the gaps in flood monitoring impact assessment and forecast modelling. *In : Bulletin of the National Natural Resources Management system,* NNRMS (B) -28 (Eds. V. Jayaraman, R.R. Navalgund, S.K. Bhan, S.K. Shivakumar). Department of Space, Government of India, Bangalore, pp. 44 -51.

Definition taken from Encyclopedia Britannica. Encyclopedia Britannica Inc. Pvt. Company. UK.

EM-DAT. The OFED/CRED International Disaster Database, www.cred.be/ emdat_universite Catholique de Louvain, Brussels Belgium.

Frater, H. (1998). *Natural Disasters : Cause, Course, Effect, Simulation.* Electronic Media, Verlag, Berlin, Germany.

Ganesh Raj, K. and Nijagunappa, R. (2004). Major lineaments of Karnataka State and their relation to seismicity : A remote sensing based analysis . *J. Geological Soc. India,* 63 : 430-439.

Ganesh Raj, K. and Nijagunappa, R. (2003). Lineaments and seismicity : An overview. *Seismicity of India,* 5 : 1-6.

Gaur, V. K. (2003). Mitigating adverse impacts of natural hazards ; The crucial role of space technologies. *In : Bulletin of the National Natural Resources Management System,* NNRMS (B) - 28 (Eds. V. Jayaraman, R.R. Navalgund, S.K. Bhan, S.K. Shivakumar). Department of Space, Government of India, Bangalore, pp. 12-23.

GOI (2004). *Disaster Management in India : A Status Report.* NDM Division, Ministry of Home Affairs, Govt. of India, New Delhi, pp. 1-86.

GOI (2005). *Disaster Risk Reduction : The Indian Model.* NDM Division, Ministry of Home Affairs, Govt. of India, New Delhi.

Government of India, Ministry of Agriculture, Department of Agriculture Co-operation, 1999-2000, 2002-2003, 2003-2004, *Annual Reports,* New Delhi.

Government of India, Ministry of Agriculture, Department of Agriculture of Co-operation, *Drought* 2002, States' Report ( Part 2- Vol. I), New Delhi.

Hooker, J. (2008). Toll rises in China quake. *New York Times* (Retrieved on 2008-06-01).

http://www.idndr.org (A/54/135 'International co-operation to reduce the impact of the *El Nino* phenomenon).

ICID (1996). *ICID Multilingual Technical Dictionary of Irrigation and Drainage,* 58 (SI) : S22-S31. John Willey & Sons Ltd., New York.

IDNDR (1999). Scientific and Technical Committee (STC) Meeting, Canberra, Australia, 15-20 February, 1999.

IFRC (2000). World Disaster Report, cited from OFDA/CRED International Disaster Database- www.cred.be/emdat-universite Catholique de Louvain, Brussels, Belgium.

IMD (1999). Cyclones and depression in 1998 in the Bay of Bengal and Arabian Sea. *Mausam,* 50 : 3.

Jai Krishana (1992). Seismic zoning maps of India. *In : Bulletin of the National Natural Resources Management Systems,* NNRMS (B) - 28, (Eds. V. Jayaranan, R.R. Navalgund, S.K. Bhan, S.K. Shivakumar). Department of Space, Govt. of India, Bangalore, pp. 17-23.

Krishnan, A. (1996). A study of the recent occurrence of widespread drought in India and their effects on country's agricultural production. *In : Proc. All India Seminar on Natural Disaster - Causes and Management* (Eds. H. Chandra Shekhar, K.S. Vijay Kumar and G.S. Srinivasa Reddy). Drought Monitoring Cell, Bangalore, pp. 7-21.

Kulshrestha, S.M. (1997). Drought Management in India. Joint COLA/ CARE Technical Report No. 1. Institute of Global Environment and Society, USA.

Kumar, S. and Parmar, A. (2005). Vulnerability and hazards reduction strategies in India. *In : Environment & Development : Challenges and Opportunities* (Ed. Jagbir Singh). I.K. International Publishing House Pvt. Ltd., New Delhi, pp. 375-392.

Mahlawat, Manjeet and Singh, Mahadevi (2003). Impact of drought and its management. Chapter 3, *In : Current Environmental Issues* (Eds.

B.B.S. Kapoor, Ahmed Ali, K.K. Singh and Chandrakanta). Madhu Publications, Bikaner, p. 37-52.

Mirza, M.M.Q., Warrick R.A., Eriksen, N.J. and Kenney, GJ. (2001). Are floods getting worse in the Ganges, Brahmputra and Meghana basin? *Environmental Hazards,* 3 : 37-48.

Negi, J.G., Agarwal, P.K. and Panday, O.P. (1998). Deformation of the Indian sub-continent due to Indian ocean floor spreading process. *The Hindu,* September 18, 1988.

NNRMS (2003). A Special Issue on Disaster Management. *Bulletin of the National Natural Resources Management System,* NNRMS (B) - 28. Department of Space, Government of India, Bangalore, pp. 108.

Ramzy, Austin (2008). The China Quake's homeless victims. Yahoo ! News, 2008-05-18 (Retrieved on 2008-05-20).

Russell, G. (2005). The humanitarian relief supply chain - Analysis of the 2004 South-east Asia earthquake Tsunami. Jarrod Goentzel, Ph.D. Massachusetts Institutes of Technology, Cambridge, USA.

Sahni, P. and Madhavi, M.A. (2003) [Eds.]. *Disaster Risk Reduction in South Asia.* Prentice Hall of India, New Delhi.

Santra, S.C. (2001) Natural Disasters. Chapter 30, *In : Environmental Science.* New Central Book Agency Pvt. Ltd., Kolkata, pp. 555-570.

Satendra and Gupta, A. (2003). Managing disaster must for sustainable development. *Employment News,* 28 (30) : 173.

Sharma, V.K. (2002). Disaster Management : A future vision. *Employment News,* 27 (24): 1-2.

Singh, H.P., Ramakrishna, Y. and Rao, G. G. S. N. (2003). Agricultural drought in India : Issues and perspectives. *In : Bulletin of the National Natural Resources management system,* NNRSM (B)-28. Department of space, Government of India, Bangalore, pp. 24-29.

Singh, K.K., Tomer, Alka, Singh, Vinod and Singh, Mahadevi (2007). Watershed development for land and water management in arid ecosystem. *Chapter 5, In: Environmental Degradation and Protection,* Vol I, (Eds. K.K. Singh *et al.*). M.D. Publications Pvt. Ltd., New Delhi, pp. 144-187.

Singh, Mahadevi and Mahlawat, Manjeet (2001). Women's participation in watershed Development. *Environment & People,* 8(6) : 13-29.

Singh, V.K. and Mahato, L.L. (2007). Career in disaster management and mitigation. *Employment News,* 32(24): 1 & 48.

Sinha, A. (2001). Disaster Management : Lessons Drawn and Strategies for Future. Theme paper of Annual conference organised by Indian Institute of Public Administration. Indra Prastha Estate, Ring Road, New Delhi, p. 42.

Sinha, A. (2003). Disaster Management : A vision for the Future. *In : Bulletin of the National Natural Resources Management System,* NNRMS (B) - 20 (Eds. V. Jayaraman, R.R. Navalgund, S.K. Bhan, S.K. Shivakumar). Department of Space, Govt. of India, Banglore, pp. 9-11.

Sinha, K.K. (1993). Isoseismal Studies, 49-58. *In : Bihar Nepal Earthquake* Aug, 20, 1988. *Geological Survey of India* (Special Publication), 31: 104.

Sinha, P.C. (1998) [Ed.]. Introduction to Disaster Management. *In : Encyclopedia of Disaster Management,* Vol. I. Anmol Publications Pvt. Ltd., New Delhi.

Smith, K. (1996). *Environmental Hazards.* Routledge, London, U.K.

Srivastava, N.N. and Ramchandran, K. (1985). A new catalogue of earthquakes for peninsular India during 1839-1900. *Mausam,* 36 (30): 351-358.

Tandon, A.N. (1992). Seismology in India : An overview upto 1970. *In : Seismology in India - An Overview* (Eds. S. Ramaseshan and H.K. Gupta). *Current Science* (Special Issue), 62 ( 1 & 2): 9-16.

Tenth Five Year Plan Document (2002-2007). Disaster Management - The Development Perspective : An Extract of the chapter, pp. 1-18.

The Australian (2008). 80,000 dead in one Burma province. The Australian, May 08, 2008.

Turner, B.A. (1970). *Manmade Disasters.* Wykenham Publications, London, UK.

Turner, Barry A. (1976). The Organizational and Inter organizational Development of Disaster. *Administrative Science Quarterly,* 21 : 378-397.

UN (1997). *Floods : People at Risk, Strategies for Prevention.* Department of Humanitarian Affairs, Geneva, United Nations.

UNDP (2004). Reducing disaster risk : A challenge for development. United Nations Development Programme. Bureau for Crisis Prevention and Recovery, New York.

UNDRO (1987). Somalia drought, May 1987. UNDRO Information Report No. 1, May 5, 1987.

Verma, R.K. (1996). Droughts in India : The Global connections and scope of prediction. *In : Proc. All India Seminar on Natural Disaster-Causes and Management* (Eds. H. Chandrashekhar, K.S. Vijay Kumar and G.S. Srinivasa Reddy). Drought Monitoring Cell, Bangalore, pp. 23-34.

WMO (1974). *Flood Problem and Management in South Asia* (Eds. M. Monirul Qader Mirza, Ajaya Dixit and Ainum Nishat). Kluwer Academic Publishers, Dordrecht, The Netherlands.

World Bank (2000). Managing Economic Crisis and Natural Disasters. *In: World Development Report 2000/2001.* Washington, DC, USA.

World Disaster Report (1998). International Federation of Red Cross and Red Crescent Societies, OUP.

World Disaster Report (2001). Environment Monitoring Data. CRED, University of Louvain, Belgium, pp. 164 -184.

## Chapter 2

# Natural Disasters—The Phenomena, Preparedness and Mitigation

*M.E.A. Mondal*[1] *and M. Raza*[2]

## INTRODUCTION

The Earth, along with other members of the solar system, is believed to have formed 4.6 billion years ago from a rotating cold gas cloud. The gas cloud was probably shocked into collapsing by a supernova explosion. As the gas cloud contracted, heat was generated because of gravitational compression, and the gas cloud started to rotate rapidly to conserve angular momentum. The central part of the cloud became hot and formed the sun, and the rapid rotation of the cloud forced some of the matter of the original cloud to form a disc rotating around the central body (sun), forming different planets (our planet Earth being one of them) (Emiliani, 1992). Earth was, initially, possibly unlayered. Through time, it has acquired different layering as a result of differentiation of constituent elements. Differentiation, in turn, required heat. The sources of this heat are mostly from the heat that was generated during the earliest part of its origin, and also the heat generated by radioactive decay of some elements through time. The dynamic machine of the earth is mainly fueled by this heat. The internal heat of the earth is, ultimately, responsible for the dynamics and kinetics of this planet. Some of the vivid manifestations of the restlessness of the earth include volcanic eruptions, earthquakes and movement of continents/plates. In addition to these endogenic processes, which are somehow related to processes operating within the earth at some depth, the surface of the earth is directly interacting with the solar energy. The continuous supply of the solar energy is mainly responsible for many exogenic processes like rainfall, floods, droughts, wind systems, storms, ocean current systems, lightning, formation and melting of glaciers etc. These endogenic and exogenic processes have been in operation from the

[1-2] Department of Geology, Aligarh Muslim University, Aligarh - 202002, E-mail: erfan.mondal@gmail.com

very inception of the earth system, and will continue to do so in future also, although not necessarily with the same pace (Montgomery, 1988). Earth scientists have studied these processes that have been operating on earth and have, mostly, understood the working principle and evolution of the earth system. But given the enormity, complex, dynamic, unconstrained, interdependent and secular nature of the earth system, there still remains many grey areas which are in the domain of active research today. Perhaps it is the most incomprehensible thing to comprehend the earth system.

The present earth system has evolved through geologic time by interaction of asthenosphere, lithosphere, hydrosphere, atmosphere and biosphere. In the geologic past, the earth system has shown its fury, and its history has been punctuated by sudden, violent catastrophic events repeatedly. Although such events continue to happen today, the reasons of such natural events are not always natural alone in modern days. In many cases, man's interference and tendency to tame the nature are linked to many natural events (Valdiya, 1987; Montgomery, 2006).

A natural event/hazard is a natural process that occurs suddenly and swiftly, taking human beings by surprise. The important natural hazards are earthquakes, tsunamis, floods, cyclones, volcanic eruptions, landslides, lightning, forest fire, bolide impact and limnic eruptions. A natural hazard does not automatically qualify for natural disaster. A natural disaster is the consequence of a natural hazard that moves from potential phase to active phase and affects human life, structures and institutions. Thus, whether a natural hazard is a disaster or not depends ultimately on the location where it occurs, and the capacity of the nation and its people to support and/or resist the event, and their resilience. A disaster occurs when it meets vulnerability. Natural hazards will never result in natural disaster in areas without vulnerability. A strong earthquake will never result in a disaster if it occurs in unpopulated areas. This may be one of the reasons that the association of the term 'natural' with 'disaster' is often disputed, as disasters are never possible without human involvement. In every phase and aspect of a disaster, it is the vulnerability, preparedness, response and resilience that matter the most.

Natural events are very much integral part of the earth system. So they cannot be prevented/stopped from occurring. But awareness, mitigation and preparedness, based on the appraisal and analysis of the risk and vulnerability in the past, and those anticipated based on the state-of-the-art research and technological advancement, in many cases, can reduce the severity of the disaster and loss to property and life. We have no other option but to learn to live with the restless earth, adopting preparedness and mitigation measures. Preparedness refers to activities or measures taken in anticipation of a disaster to make sure that effective and appropriate actions are employed, whereas mitigation refers to measures taken in order to minimize the effects of the disaster and its prevention.

The Indian subcontinent is very much prone to many natural hazards like floods, cyclones, landslides, earthquakes and tsunamis because of its highly diversified natural features like the mighty Himalayas, long coast line, vast river plains, desert, plateaus and also because of the dynamic nature of the Indian plate (Valdiya, 1987). The plate is still in motion and pushing the Russian plate in the north. Stress accumulated because of this motion is often released in the form of earthquakes. Long coast line is prone to tropical cyclones arising in the Bay of Bengal and the Arabian Sea. Landslides are common features in the Himalayas and the Nilgiris. Floods often devastate many river plains. Here, we give an account of the causes, preparedness and mitigation measures of naturals disasters with special reference to Indian subcontinent.

**Earthquakes**

***The Phenomenon***

Earthquake is ground shaking, caused by sudden release of strain energy, accumulated over long time interval. It originates in a limited area at some depth due to the disturbance of the elastic equilibrium of the rock. Various types of seismic waves originate and pass through the interior, and also through the surface of the earth, shaking the ground that we feel as earthquake. Earthquakes can be broadly classified into three groups: (i) Interplate earthquake, occurring at the edges of the interacting plates, e.g., Himalayan seismic belt (Raval, 2000) (ii) Intraplate earthquake, occurring within a continental or oceanic plate and (iii) Stable Continental Region (SCR) earthquake (Rajendran, 2000), occurring in cratonic regions, e.g., Latur (Killari), 1993, Bhuj, 1819 (Rajendran and Rajendran, 2 001) and Bhuj, 2001 earthquakes. Two models are widely applied to explain the origin of the earthquake: *Elastic Rebound model* and *Dilatancy model*. In Elastic Rebound model it is envisaged that rocks under stress deform elastically till they fail. After failure, the rocks snap back to their original unstressed condition. When the rocks suddenly fail along a fracture, called fault, earthquakes occur. In dilatancy model, it is considered that rocks under stress dilate and open up the pores and cracks. With continued stress, the cracking becomes extensive and water seeps through the cracks and pores, resulting in increase of fluid pressure, and effectively lubricating the rocks and, thereby, decreasing the frictional resistance. As a result, rocks eventually slip to release buildup stress. In addition to these two models, Induced Seismicity (Bell, 1980) has also been invoked for many earthquakes, e.g., Koyna earthquakes. In this hypothesis, it is thought that increase in pore pressure induced by reservoir reduces the strength of rocks leading to failure along faults. Besides the direct effects of earthquake devastations, there are many earthquake-related hazards like landslides, tsunamis and soil liquefaction. Soil liquefaction occurs in a buried layer of unconsolidated water saturated soil when it is subjected to jarring by an earthquake. The liquefied soil looses its strength and causes the buildings to topple.

Can we control earthquakes? Scientists have long contemplated to conduct nuclear explosions in meticulously chosen sites where stress is being accumulated, e.g., known active faults (like San Andres Fault, Allah Bund Fault in Gujarat) to unstick locked segments, causing small earthquakes and, thus, preventing future large earthquakes. Fluid injection along major faults has also been considered. Injected fluid will decrease the resistance to shearing and allow the rocks to slip. This will result in small harmless earthquakes, as large amount of stress could not be accumulated in the rocks. But given the dynamic nature of the earth system, there is no guarantee that only small earthquakes will be generated; this can also lead to the release of all the stress of the rocks at once, causing devastating earthquakes. In case of nuclear explosion, there is always great concern about radiations. Keeping these things in mind, such proposals have never been attempted (Montgomery, 1988).

### *Prediction*

If we cannot control earthquakes, can we predict it? Although much is known today where earthquakes are likely to occur, there is currently no scientific way to predict the months and days when an earthquake will strike at a given location. A scientifically valid prediction of a natural event should include: period range within which the event will occur, location of the event, range of magnitude and probability of the event to occur. Earthquake prediction still remains an elusive goal and is, probably, the biggest challenge of this century. Although many earthquake precursor phenomena like micro seismicity, unusual animal behaviour and emission of radon have been put forward, none of them could be universally applied (Agrawal, 2000). One of the classical scientific predictions comes from the Haicheng earthquake of magnitude 7.3 on February 4, 1975 in China. This prediction was done mainly from the study of Rn emission in a hot spring.

***Preparedness and Mitigation***

The most important measure to cope up with seismic hazards is preparation of seismic zoning maps. The zones are drawn based on seismic evidence obtained instrumentally, and also from historical record. A seismic zoning map shows the zones of seismic vulnerability of a particular region, and provides a broad scenario of the seismic risk. The seismic zone maps are revised from time to time as more data pour in, and as understanding of seismotectonics of a particular region is built up. A glaring example of this is the elevation of seismic zone status of Killari (Latur) form zone I to zone III after the devastating earthquake in 1993. Seismic zones should be supplemented by microzonation maps with much more technical informations. Seismic hazard maps of India are also being prepared under the Global Seismic Hazard Assessment Programme (GSHAP) (Giardini, 1999; Giardini and Basham, 1993). Based on major tectonic features and spatial distribution of seismicity, seismic hazard map for India has been prepared under GSHAP, and 86 potential seismic source zones have been delineated. Major earthquake research programmes have been undertaken after the Latur earthquake. Since record of historical earthquakes is, in many cases, insufficient because of long recurrence interval of the earthquakes, there is great need to undertake research of prehistoric earthquakes (paleoseismology) in a big way. With the advancement of the Quaternary dating techniques, it is now possible to date the Quaternary events more precisely and accurately. Earthquake mitigation measures include construction of earthquake resistant buildings. Public awareness should be created and strict earthquake-resistant building construction rules be formulated and adhered to. Seismically most active area should be barred for future and further development. Public warning and disaster management system should be streamlined, focused and effective.

The Indian subcontinent with a variety of geotectonic features is very much prone to earthquakes. The Himalayan belt is one of the highly seismically active regions. Seismicity of varying magnitudes is exhibited in different places in fault ridden peninsular India. Notable among them are Latur, Bhuj and Koyana. Earthquakes that occurred in the Indian subcontinent in the 21$^{st}$ century are: Bhuj earthquake (Jan, 2001), Andaman Nicobar (Dec. 2004) earthquake and Muzafarabad/J. & K. (Pakistan and

India) earthquake (Oct. 2005). There has been no major earthquake after 1950 in Assam region. There also appears a "locked" segment in the western part of the Nepal Himalaya from where major earthquakes have not been reported. Strain energy is being accumulated, and these places are the probable places for future earthquakes. Seismic hazard zonation studies are actively being pursued in the country, which may help in better preparedness and mitigation measures.

## Tsunami

### *The Phenomenon*

Tsunamis are large destructive sea waves having very large wavelengths to the tune of hundreds of km, and very low amplitude of around one metre in open sea. Because of such long wave lengths, tsunami waves behave like shallow-water waves. These waves are generated mainly in three major ways. Most common reason is undersea earthquake, whereby a fault is generated which may vertically displace the sea floor. The second reason may be due to submarine volcanic eruption. The crater and/or the flank of the volcano may abruptly change its height either by upliftment or collapse of the vent. Third major cause may be a submarine landslide. Whatever may be the reason, all these processes cause the sea floor to deform abruptly, resulting in large scale depression or upliftment of the ocean floor, disturbing the equilibrium of the overlying water surface. In case of depression of the ocean floor, water flows in from all sides, exhibiting withdrawal of the sea from nearby coasts. If the displacement of the ocean floor is upward, water flows outwards in all direction. When the tsunami waves reach the coast, their amplitudes increase to heights more than tens of metres, causing havoc to life and property.

### *Preparedness and Mitigation*

Loss to life and property due to tsunami can be minimized by proper preparedness and mitigation measures. Some of the preparedness measures include mapping of tsunami prone areas, delineation of most vulnerable areas, development of early warning system, planning of evacuation routes and management, so that effective evacuation is possible within short notice. Possible risk reduction measures should emphasize on construction of barriers, breakwaters and sea walls along tsunami prone coastlines.

Records show that the Indian coast was struck by tsunami on Nov. 27,1945 when the western coast around Mumbai was struck. The tsunami originated near Makran coast and reached up to Lankan coast. After that there was a long lull till Dec 26, 2004 when the earthquake in Sumatra-Andaman generated an intense tsunami that struck the eastern coast of India (Roza, 2007). Despite a gap of several hours between the occurrence of the earthquake at Andaman and the striking of the tsunami at the Indian coast, there was no tsunami early-warning system at or near the countries that were affected. Tsunami detection in open sea is not easy, as in deep water it has very little height. A network of sensors is needed to detect the tsunami at the early stage. Even a bigger problem, particularly in countries like India, is communication network to issue timely warning, and evacuation mechanism and sensitization of populace living along the tsunami prone coast line. After the aftermath of the Dec 26, 2004 tsunami, there is some awareness among the common people. It is also much more surprising that international organizations concerned with various kinds of natural disasters did not feel the importance about monitoring of tsunami in the Bay of Bengal and the Indian ocean. It is generally perceived that the recent tsunami disaster could have been avoided, or its severity minimized, had the concerned international agency perceived such thing. The United Nations has, however, started working on tsunami warning system in the Indian ocean.

## Landslides

### *The Phenomenon*

Downslope movement of rock, soil or any earth material under the influence of gravity is generally referred to as mass wasting or mass movement. Landslide is a term generally reserved for rapid mass movement in hilly and mountainous regions. The material involved may range from solid rock to mud, and all possible combinations between the two. Landslides occur whenever the downward pull of gravity is overcome by the sliding forces. The important causative factors are steep slope, excessive rainfall and deforestation. Severe storms, earthquakes, volcanic activity, coastal wave attack, frosting and thawing, presence of expansive and quick clay can cause widespread slope instability resulting in setting of mass movement (Montgomery, 1988, 2006; Valdiya, 1987). The movement may range from sudden fall, through slippage to extremely slow movement, known as creep. Landslides can be grouped into following categories : *Rock Falls*, Slumps, Avalanches and Flows, Lahars and Creep. *Rock Fall* refers to free-falling action where the rock is not always in contact with ground below. Such phenomenon is very common near a vertical slope or cliff, where the under support gets eroded. *Slump* mostly occurs in soil, and is often characterized by rotational movements. An *avalanche* is a hazard that involves slide of a large snow mass down a mountain side. This is a major danger faced by people in mountains in winters. *Lahar* is a natural disaster closely associated with volcanic eruption and involves a large amount of material, including mud and ash, sliding down the side of a volcano at a rapid pace. One of the infamous examples of lahar disaster is the Tangiwai disaster which killed an estimated 23,000 people in Armero Columbia in 1985. *Mudflows* or *mudslides* are special case of landslides, where excessive rainfall causes loose soil on steep slope to flow or slide. *Creep* is imperceptive slow movement which cannot be detected *easily* (Heimsath *et al.* 2002). Presence of curved tree stems, tilted monuments or *telephone/electricity* poles on a slope is often indicative of creep.

### *Preparedness and Mitigation*

To address landslide hazard, first step is identification of landslide prone areas and preparation of zoning/micro-zoning maps (Glade and Crozier, 2005). This is done on the basis of geomorphic features, palaeolandslide, geology, topography, climate, groundwater and land use pattern. Landslide can be a serious earthquake-related hazard, as earthquake may trigger sliding of unstable slopes. Mitigation measures may include slope modification, excavation and filling, improving drainage, forestation of slope, supporting or anchoring the existing landslide prone slope and application of geotextiles. In order to quantitatively assess and predict the extent of mass movement, it is necessary to model the process of the mass movement and analyze the slope stability (Jones, 2004). Slope stability study should include rock mass characterization, permeability discontinuity characterization, groundwater observations and stress condition.

Landslides are very common in the young fold mountain belt of the Himalayan region. Other regions like Western Ghats, Eastern Ghats and Nilgiri hills also experience landslides, but with lesser frequency and intensity. In India, landslide studies, including rock mass characterization and landuse zonation, are being carried out in a big way.

## Flood

### *The Phenomenon*

Streams tend to maintain a dynamic equilibrium among factors like slope, sediment load and discharge rate. The streams are active erosional and depositional agents, and continually adjust their channels to accommodate the average annual discharge. For major part of the year, water flows through the river well below or up to the river bank height. The river is at flood stage when water level rises above the bank heights and overtops it. Floods should not be considered unnatural or unusual events. Instead, floods are integral part of the river dynamics and absolutely normal phenomenon. Many factors like climate, topography, soil type and vegetation together determine the severity of floods (Subramanya, 2003). One of the important reasons of flooding may as well be neotectonics. Neotectonic activity can disturb the equilibrium profile of the river and, as a result, the river may be bonded and during excessive rainfall may get easily flooded. The reason of the frequent floods of the rivers Ghagra, Gandak, Naraini and Kosi may be ascribed to neotectonics. The channels of these rivers and their associated floodplains are not higher than the floodplain of the river Ganga to which they meet, i.e., the channels of the tributaries and the main river are almost at same level. As a result, during rainy seasons when there is excess rainfall, water of these tributaries does not get drained out easily and quickly through the main river, Ganga. So these rivers become wild during floods. The disturbance in the equilibrium profile of the rivers is due to the upliftment of the southern track of the river Ganga, the Chotanagpur plateau (Valdiya, 1984). The upliftment, though slow, imperceptible and gentle, of the Chotanagpur plateau is manifested by prominent geomorphic features like river terraces, water falls and mild seismicity.

### *Preparedness and Mitigation*

Floods are, to some extent, predictable. So by taking flood control measures its severity can be minimized. Human activities unintentionally aggravate the flood hazards by encroaching the floodplain, constricting the river channel, and draining runoff quickly via sewer line. Active flood plain of rivers should be left as it is, and any construction or cultivation along the floodplain should be avoided. Preparation of flood zoning maps of vulnerable region is of utmost importance. Other strategies to mitigate flood damage may include construction of embankments, retention ponds, check dams and canalization of excess water (Ghosh, 1997).

The Indo-Gangetic plain, which supports about 40% of the country's population, is highly prone to flood. Other areas where severe flooding occurs are Brahmaputra basin and Narmada, Godavari and Mahanadi valleys. In recent years, the desert area in the western part is also experiencing floods, although localized in extent. Undeveloped natural drainage system, and presence of kankar horizon at shallow depth, which prevents water percolation are important reasons for this.

## Cyclone

### *The Phenomenon*

Global wind system conforms to a pattern generated by temperature differences between equator and the poles, coriolis effect and the differential heating of ocean and continents. Over the tropical region, air gets abnormally heated and rises quickly. As the air rises, it rotates and creates an area of low pressure. As pressure falls, wind speed increases and cloud band starts spiraling around a centre. With further fall in pressure, the surrounding warm air gushes in and causes it to spin faster and faster. These strong currents create a spirally, moving vortex of wind called cyclones. Where the vortex touches the ground, it creates concentrated destruction. The cyclone, then, moves towards low pressure landmass causing high storm and heavy rainfall. These storms are known differently in different regions. In Indian subcontinent these are known as cyclones, in North and South America as hurricanes, in China and Japan as typhoons, and in Australia as Willy-Willy.

### *Preparedness and Mitigation*

Most important pre-disaster preparedness is to delineate and prepare cyclone prone areas. Strengthening meteorological research to monitor and track the movement of cyclones (Holland, 1993), proper warning system, evacuation plan and construction of cyclone resistant buildings constitute essential components of preparedness and mitigation of cyclones (Joseph, 1994).

The eastern coast of India from West Bengal to Tamil Nadu, through Orissa and Andhra Pradesh, is highly vulnerable to cyclones, accompanied by heavy rainfall. In the western coast, northern Gujarat and Saurashtra face this nature's fury mostly in post-monsoon period.

## Volcanoes

### *The Phenomenon*

Volcanoes are the vents in the surface of the earth through which magma and associated gases come out. An erupting volcano is, indeed, the most awesome and terrifying sight in nature. Nature of erupting material, and character of volcanoes, are directly related to the magma composition and the tectonic setting in which the volcanoes erupt. Based on activity, volcanoes can be grouped as Active, Dormant and Dead. There are-thousands of active volcanoes across the globe. *Active* volcanoes are those which have erupted in recent history. *Dormant* volcanoes have not erupted recently but appear fresh in look, without much weathering. Dead volcanoes have not erupted for quite a longtime and exhibit deep erosion.

The reason of volcanism is the generation of magma by partial melting of rocks at some depth inside the earth. Generation of magma is a complex phenomenon, and is a function of tectonics, geothermal gradient and pressure (lithostatic and/or fluid pressure). Magma which is formed, being lighter in density than the surrounding rocks, ascends up and may erupt on the surface of the earth as volcano in the temperature range of 650 to 1200 °C. The material which is ejected out of a volcano is hot molten lava, ash, gas and a variety of other rock fragments. Hazards due to volcanoes are related to lava flows, pyroclastics ash falls, lahars and emission of toxic gases.

***Preparedness and Mitigation***

Measures to cope with the volcanic hazards are zonation of volcano prone areas, prediction and early warning system. Advanced precursor phenomena of volcanism include microseisms, uplift, bulging and tilt of volcano surface, emergence of thermal spring, increase of heat flow around the volcanoes etc. Civil defence measures include diversion of lava flows by construction of heavy protective walls or trenches.

India is almost free from this natural hazard, except for the Barren Island which got activated recently, ejecting out toxic gases and ash during 1995-1996.

**Lightning**

Lightning is characterized by discharge of electricity between rain clouds, or between rain clouds and the earth. It is manifested in the form of extremely bright arc of light accompanied by roar of thunder. The lightning causes the air to get heated and expanded. The expanded air collides with the cooler air, creating a roar of explosion. Formation of lightning is a complex phenomenon. In most cumulonimbus clouds, the bottom of the cloud gets negatively charged, whereas the top portion of the cloud accumulates positive charges. When the electrical potential in the clouds reaches about ten thousand volts, ionization occurs along narrow paths, resulting in a flash of light called lightning (Uman and Rakov, 2003). The negatively charged particles descend from the bottom of the clouds to the ground.

It is most important to observe some safety measures to avoid lightning. During rains, we should avoid standing under tall trees or objects. We should also avoid large water bodies and metal objects. Tall buildings should be fitted with lighting rods.

**Limnic Eruption**

Limnic eruption, also called lake overturn, is rather a rare natural disaster where $CO_2$ suddenly erupts from deeper parts of the lake, causing suffocation. Till date, only two limnic eruptions have been observed and recorded. A limnic eruption occurred in 1984 in Cameroon in Lake Monoun, causing death of 37 people residing nearby. The other limnic eruption, much larger in size, occurred in 1986 at the nearby Lake Nyos, killing about 1800 people.

**Bolide Impact**

The earth experiences impact of astronomical bodies called bolides, like meteors or asteroids of varying sizes, from time to time. There have been numerous meteorite impacts on the earth in the geological past. They have played important role in the evolution of earth in its long journey. It is also believed that meteorites have been responsible for mass extinction of some species. There has been no great bolide impact in recent past, except for three impacts that occurred during the last century. In 1908, a bolide impact occurred in Siberia and devastated a large area. Had it fallen on a populated area, it would have destroyed the entire city. Other strike was in 1947 in Brazil, and the third one struck British Columbia in 1956. Fortunately, all the impacts were in unpopulated areas and none was harmed. Earth may face the risk of being struck by big impacts in future also. Scientist now contemplate to strike bigger asteroids posing threat to earth by nuclear warhead in the space so that the asteroid may be deflected, or broken into smaller pieces.

## CONCLUSION

Although natural disasters cannot be prevented, proper disaster mitigation and preparedness can reduce the damage. The only way left is to identify vulnerable areas before the disaster, and take appropriate steps to protect against the disaster and minimize the damage. In addition to scientific and technical inputs regarding a natural disaster, keeping in view the unprecedented nature of disasters, there is urgent need to strengthen awareness and perfect coordination amongst policy makers, decision makers, administrators, professionals and financial institutions. If required, appropriate legislation and amendments be done to strengthen enforcement mechanism. It is imperative on us to learn the delicacy of the Earth System, which has evolved through geological time and will continue to evolve in future also. Disaster preparedness and mitigation measures can only be effective in the long run if we achieve a close understanding of the planet Earth and its dynamic system.

## REFERENCES


Agrawal, P.N. (2000). Seismological aspects of earthquake damage reduction. *Sixth IGC Foundation Lecture, IGC-2002 Udaipur.*

Bell, F.G. (1980). *Engineering Geology and Geotechniques.* New-nes-Butterworths, London.

Emiliani, C. (1992). *Planet Earth.* Cambridge University Press, U.K.

Giardini, D. (1999). The Global Seismic Hazard Assessment Program (GSHAP) 1992-1999. Annali Di Geofisica, 42 (6): 957-974.

Giardini, D. and Basham, P. (1993). Global Seismic Hazard Assessment Program for the UN/IDNDR, *Ann. Geofis.*, 36(3-4): 3-13.

Ghosh, S.N. (1997). *Flood Control and Drainage Engineering.* Oxford & IBH Publication, New Delhi.

Glade, T. and Crozier, M.J. (2005). *Landslide Hazard and Risk.* John Wiley & Sons, New York.

Heimsath, A.M., Chappell, J., Spooner, N.A., Questiaux, D.G. (2002). Creeping soil. *Geology,* 30: 111-114.

Holland, G.J. (Ed.) (1993). *Global Guide to Tropical Cyclone Forecasting.* World Meteorological Organization, WMO/TD No. 560, Report No. TCP-31.

Jones, D.K.C. (2004). *Landslide Risk Assessment.* Thomas Telford, London, UK.

Joseph, P.V. (1994). Tropical Cyclone Hazard and Warning and Disaster mitigation Systems in India. *Sadhana Academy Proc. Engineering Science.* Indian Academy of Sciences, Bangalore, 1955-56.

Montgomery, C. W. (1988). *Physical Geology,* Wm. C. Brown Publishing Co., Dubuque., IA, USA.

Montogomery, C. (2006). *Environmental Geology*. McGraw-Hill Publishing Company, New Delhi.

Rajendran, C.P. (2000) Using geological data for earthquake studies: a perspective from peninsular India. *Current Science*, 79 : 1251-1258.

Rajendran, C.P. and Rajendran, K.F. (2001). Characteristics of deformaticn and past seismicity associated with there Kutch Earthquake, northwestern India. *Bull. Seis. Soc. Amer.*, 91: 407-426.

Raval, U. (2000). Laterally heterogeneous seismic vulnerability of the Himalayan arc: a consequence of cratonic and mobile nature of underthrusting Indian crust. *Current Science*, 78:546-549.

Roza, G. (2007). *The Indian Ocean Tsunami*. The Rosen Publishing Group, Malden, Massachusetts, USA.

Subramanya, K. (2003). *Engineering Hydrogeology*. Tata McGraw-Hill, Publishing Company, New Delhi.

Uman, M. A. and Rakov, V.A. (2003). *Lightning: Physics and Effects*. Cambridge University Press, Cambridge, UK.

Valdiya, K.S. (1987). *Environmental Geology: Indian Context*. Tata McGraw-Hill, Publishing Company, New Delhi.

Valdiya, K.S. (1984). *Aspects of Tectonics*. Tata McGraw-Hill, Publishing Company, New Delhi..

# Chapter 3

# Landslide Hazard Mitigation and Management in the Himalayan Region

*Jitender Saroha*

## INTRODUCTION

India is considered as one of the world's most disaster prone nations. Earthquakes, floods, droughts, cyclones and landslides are the major frequent disasters in India. Landslides due to characteristic physiographical, geological, climatic and tectonic conditions affect various regions of India differently. Landslides are a major problem particularly in the Himalayan region. Every year this region faces several hundreds of landslides, from Jammu and Kashmir to the North Eastern states including Himachal Pradesh, Uttarakhand and Sikkim Himalaya. In addition to natural proneness to landslides, improper planning and haphazard construction practices are also responsible for increased incidences of landslides and consequent damages (Anbalagan and Singh 1996). Landslide hazard mitigation and management is required for the sustainable development of the Himalayan region. The objectives of the present paper are - (i) to define and identify the causes of landslides, (ii) to highlight variations in occurrence of landslides in India, (iii) to identify different landslide hazard assessment techniques and, finally (iv) to highlight the event modification and vulnerability modification techniques.

### Causes of Landslides

A landslide is a major geological hazard, which poses serious threat to human beings and various other infrastructures like highways, railways and civil structures like dams, buildings and other structures. A landslide is the downward and outward movement of slope forming materials, composed of rock, soil, artificial fills (dumping), or a combination of all these, along the surface of separation by falling, sliding, flowing under

Department of Geography, B.R. Ambedkar College, University of Delhi, Delhi, Email: dabasjeetu@rediffmail.com, jeetu43@yahoo.com a fast or slow rate, but under the action of gravitational force, where the triggering factor may be natural or anthropogenic. The term landslide has been accepted, and is being used commonly around the world as a synonym of 'mass movement' or 'mass wastage' (Anbalagan, 2007).

All landslides are a form of slope failure. They happen when the shear stress within a slope exceeds the strength of the slope material. Then the slope fails, and millions of cubic feet of rock and soil materials can shear away from the slope and move hundreds or thousands of feet down the hill. The most important factor promoting landslides is an increase in the angle of the slope: the steeper the slope, the more prone it is to landslides. Another common factor contributing to landslides is the addition of water to the area. Water lifts or pushes the grains apart in the soil or rock, reducing the internal friction of the soil, and counteracting the gravitational forces that hold the slope in place. The repeated freezing and thawing of water in cracks can be responsible for rock falls. The process is called frost wedging, in which the expansion during freezing widens the cracks, and allows the water to penetrate deeper into the rock when the thaw occurs. Individual blocks can be wedged out of the cliff face, falling independently, or causing such a loss of cohesion that larger portions of the cliff face can collapse.

Landslides can also be caused by earth tremors. Earthquakes, volcanic eruptions, and even heavy machinery or trains passing on nearby roads or railroads, have been known to induce tremors that start landslides. Another factor that promotes landslides is the removal of lateral or basal support from a slope. In nature, this occurs because of erosion by either meandering rivers or wave actions on ocean cliffs. Vegetation changes contribute to landslides in a variety of ways (Anbalagan, 1996). In high mountain valleys, the bedrock is wedged apart by roots of trees. On gentler slopes, vegetation helps to anchor loose soil materials and prevent landslides. Wildfires have been responsible for promoting landslides by destroying tree cover; areas freshly clear cut by the logging industry, or cleared for housing have also been reported as sites of increased landslide activity.

These diverse factors causing landslides can be categorized into - (i) Natural factors and (ii) Anthropogenic factors. The natural factors are further subdivided into inherent and external factors. The inherent factors represent the inherent characteristics of hill slope, and they can be studied and evaluated on the slope itself. These factors include geology, slope gradients, local relief, hydrological conditions, land use and land cover.

The external factors are the outside factors, which cannot be studied on a hill slope. They usually affect a large area and, hence, are called regional factors. These include concentrated rainfall and earthquakes. These are also known as triggering factors, as many times they initiate landslides. Among the anthropogenic factors deforestation, improper land use and construction activities are the important ones (Fig.1).

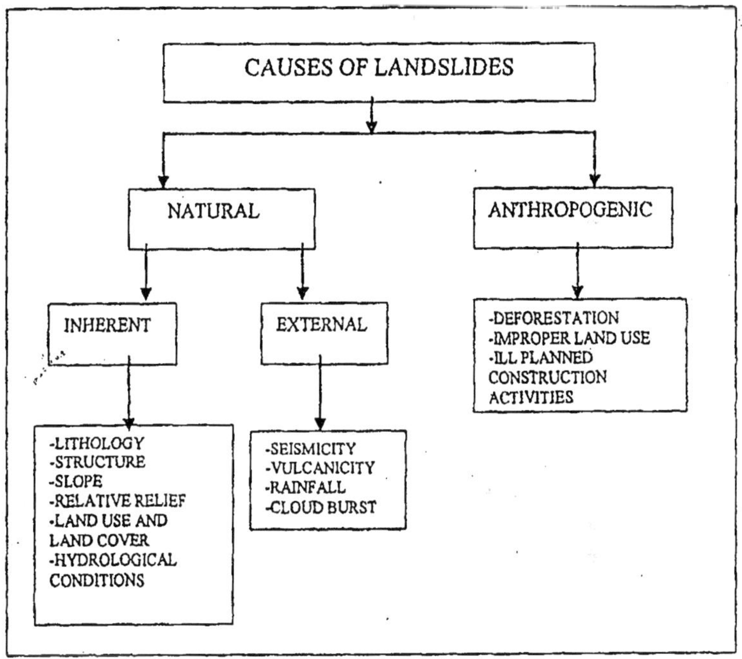


**Fig. 1: Causative Factors of Landslides (After: Anbalagan *et al.* 2007)**

## Landslide Hazard Zones of India

Landslides are among the major hydro-geological hazards that affect large parts of India, especially the Himalayas, the Northeastern hill ranges, the Western Ghats, the Nilgiris, the Eastern Ghats and the Vindhyas, in that order. In the Himalayas alone, one could find landslides of every fame, name and description - big and small, quick and creeping, ancient and new. Similarly most of the northeastern region is bristling with landslides of a bewildering variety. Then, there are landslides in the Western Ghats in the south, along the steep slopes overlooking the Konkan coast. Landslides are also very common in the Nilgiris, characterized by a lateritic cap, which is very sensitive to mass movement.

In India, the incidence of landslides in Himalayan region and other hill ranges is an annual and recurring phenomena. India, a country with varied physiographic and climatic conditions, faces the vagaries of landslides frequently. Approximately 0.49 million km$^2$ or 15% of land area of the country is vulnerable to landslide hazard. Out of this, 0.098 million km$^2$ is located in Northeastern Region, and the rest 80% is spread over Himalayas, Nilgiris, Ranchi Plateau and Eastern and Western Ghats (Fig. 2).

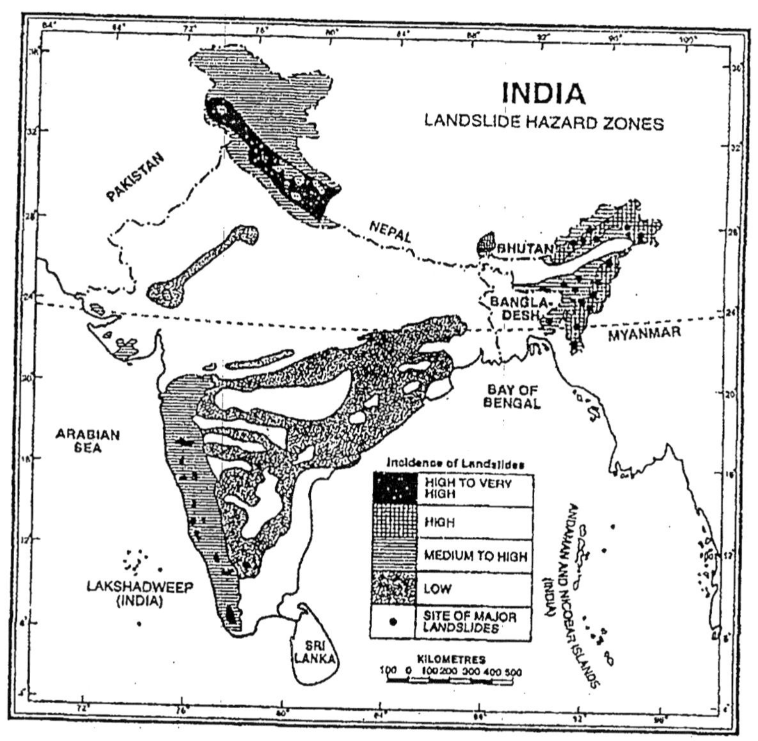


**Fig. 2: Landslide Hazard Zones of India**

The Himalayan region, being geologically young and susceptible to earthquakes and intensive soil erosion, is highly prone to occurrences of landslides. Landslide and related mass movement activities are ubiquitous here. This region is prone to landslides because of the interplay of many causal factors such as - lithology, structure, steepness of slopes, hydrological and climatic conditions, and vegetative cover and anthropogenic interventions. The frequency of landslides has further increased with road construction (Rudra *et al.*, 1982). Over the years, due to increasing cultural activities, the incidences of landslides have shown a disturbing trend of occurrence with higher damage to life and property in the Himalayan region. The recognition of the more widespread occurrence of slope instability in the Himalayan region leads, naturally, to the conclusion that the costs of slope failure will undoubtedly escalate as a consequence of development pressures unless active steps are taken towards landslide mitigation and management. Therefore, anticipatory planning based on carefully prepared landslide hazard assessments is required. There are several reasons for this view:

(i) The popular literature on hazards has tended to over-emphasize the significance of death tolls in assessing the importance of hazard impacts. Undoubtedly, a large death toll raises the profile of an event into the attention frame of decision-makers and politicians, but death toll cannot be used as a surrogate for total impact measured in economic terms, resulting from destruction, damage, disruption and delay. Massive investments in infrastructure, buildings and commercial activities indicate growing vulnerability to landslide hazard impacts.

(ii) Even minor slope failures can have extremely costly repercussions, especially the dislocation of sub-surface networks.

(iii) Landslide induced disruption to engineering projects have the potential of being politically embarrassing and damaging to reputations as well as extremely costly.

Unfortunately, landslide hazard is frequently poorly perceived and underestimated by those involved in decision-making. Despite the slowly growing awareness of the significance of the landslide hazard, the true costs of mass movements in monetary terms remain unknown. It is easy to estimate the direct costs of repair and rebuilding, but it is extremely difficult to estimate the costs of death, injury, disruption and delay. Therefore, landslide hazard mitigation and management is required for the sustainable development of the Himalayan region. To achieve this, the present study highlights following requirements - (1) the need to heighten the perception of decision-makers towards the nature, scale, distribution and causes of landslides and significance and spatial variations of landslide hazard, (2) the need to increase awareness to the range of adjustments (both structural and non-structural) that can be adopted to ameliorate the problem, (3) the need for landslide management to be incorporated as an element of developmental planning, and (4) the necessity of developing and refining rapid and meaningful landslide hazard evaluation practices.

There are huge numbers of relatively small to medium sized slope movements that, cumulatively, impose at least as great, if not greater, cost to human society as the rare catastrophic failures. The costs of these are much more widely distributed, and both their occurrence and frequency are exacerbated by human activity, so that total losses attributable to landslides are growing rapidly. The present scientific knowledge enables prediction of such non-catastrophic failures with moderate accuracy, and their impact can be effectively mitigated by stabilization measures and land use planning. They, therefore, represent the most urgent focus of landslide hazard assessments and management strategies, because of the potential for achieving significant hazard loss reduction.

**Landslide Hazard Assessment**

Landslide is considered to be one of the potentially most predictable of geological hazards and this means high potential for loss reduction. Such statements are made with particular reference to small and medium scale events, and depend upon the three basic assumptions identified by Varnes (1984):

(i) Principle of uniformitarianism- the conditions that led to slope instability in the past and the present will apply equally well in the future.

(ii) That the main conditions that cause landslide can be identified.

(iii) It is usually possible to estimate the relative significance of individual factors. This facilitates assessment of degree of hazard by examining the number of failure-inducing mechanisms present in any area.

Scientific landslide hazard assessment provides basis for landslide prediction as well as landslide forecasting. Numerous different approaches to landslide hazard assessment have been developed over the last three decades. Hansen (1984) identified three main groups of techniques:

(i) Geotechnical investigations - involving the detailed analysis of surface and sub-surface conditions and ground materials. Geotechnical investigations are performed to evaluate those geologic, seismologic and soil conditions that trigger landslides, endangering the safety of the surroundings (Sarda and Gupta, 2006).

(ii) Direct mapping - involving the analysis of landforms and identification of existing landslides so that areas of past instability can be identified, thereby facilitating extrapolation from areas of recognized past instability to similar situations which may suffer slope failure in the future. The simplest form of direct mapping is the landslide inventory map, which displays the distribution of recognized landslides. Sikkim Himalayan trouble spots (areas prone to minor and major landslides) were mapped by the Task Force on Environment and Development (Fig. 3). Another one is geomorphological map, which places landslides in their setting along with causes.

(iii) Indirect mapping which requires the collection of data on the causes and mechanisms of landslides, so that assessment of slope stability can be made by the application of known landslide- inducing parameters.

(iv) To these may be added the fourth approach of land systems mapping, which lies intermediate between direct and indirect mapping. It involves the definition of areas within which occur certain predictable combinations of surface forms, soils, vegetation and surface processes. The result is a hierarchical classification of terrain into land systems (large-scale division) composed of land facets/land units, which are further divisible into land elements.

By following direct, indirect or land systems mapping approaches, landslide inventory maps (landslide hazard maps), landslide susceptibility maps (landslide potential maps) and landslide risk maps (involving the assessment of potential losses) are prepared. The most important feature of landslide susceptibility maps and landslide risk maps is the spatial divisions of the surface into areas of different levels of threat (zonation and micro-zonation). It is these divisions that provide essential framework for land use planning, building regulations and engineering practices.

Preparation of a comprehensive landslide hazard zonation map requires intensive and sustained efforts (Anbalagan, 1992). The problem is highly interdisciplinary in nature. A large amount of data concerning many variables, covering large slope areas has to be collected, stored, sorted and evaluated. The use of aerial photographs and adoption of remote sensing techniques helps in the collection of data. For storage, retrieval and analysis, adoption of computerized techniques (Geographical Information System) is useful. The main steps in zonation are:

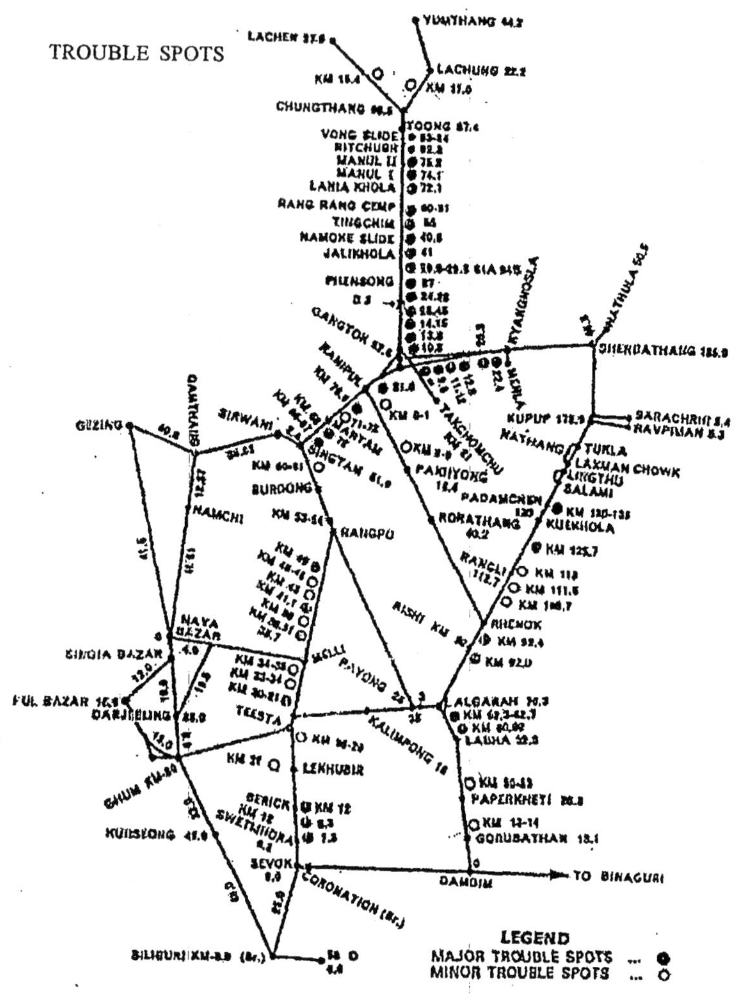


**Fig. 3: Landslide Inventory Map of Sikkim Himalaya (After: Task Force Report on Environment and Development of Sikkim, 1982)**

1. Identify the factors
2. Give weightage to factors
3. Prepare analytical map for each selected group of factors
4. Superimpose different analytical maps

5. Finally, produce a combined map, i.e., landslide susceptibility map
6. Add the degree of risk to produce landslide risk map.

These maps (LIM, LSM, LRM) have multifarious uses, some of which are listed below:

1. In the preparation of development plans for townships, dams, roads, and other development
2. General purpose Master Plans and Land use Plans.
3. Discouraging new development in hazard prone areas.
4. Choice of optimum activity pattern based on risk zones.
5. Quick decision making in rescue and relief operations.

LHZ maps provide the first hand information to planners and developers, so that landslide hazards are taken into consideration for the planned development of the area. These maps are time dependent and change in any one factor may require re-evaluation (Gupta *et al.*, 1993). Clearly such maps have a large number of users, including several Government Departments, and private agencies as well as NGOs involved in any type of development, construction of disaster management work.

**Event Modification (Stabilization)**

Loss of life and property damages from landslides usually leads to demands for engineering works to stabilize the slope. However, distinctions are made between emergency and permanent works. Emergency responses, designed to protect public safety and prevent further immediate damage, are usually undertaken satisfactorily, but government funds are only very reluctantly made available for permanent slope stabilization. This is because of lack of specialized geological information and high costs involved. If these informational and financial problems can be overcome, the stability of the slope may be improved by a variety of techniques:

(i) Geometric methods (Excavation and filling methods)- can be used to produce a more stable average slope. This type of reshaping is usually successful, but more difficult and expensive as the slide area increases. Specific techniques include the strengthening of the toe of potential landslide and the replacement of failed material with lighter loads,

(ii) Hydrological methods (Drainage, especially sub-surface drainage)

- are equally effective when changes in pore water pressure has been caused by a rise in the water table. Drainage methods range from the removal of surface water, and the drainage of tension cracks, to the insertion of trenches filled with gravel or horizontal drains. Theoretical consideration, as well as practical experience, has clearly shown that stability of saturated slopes can be improved significantly by lowering the water levels. This can be realized by installing horizontal drains or deep trench drains.. Shallow drainage work has the main function of intercepting surface runoff water and keeping it away from potentially unstable areas. In reality, on rocky hillsides this type of measure alone, although contributing to reducing the amount of infiltration, is insufficient to stabilise a hillside. Deep drainage is the most effective with this type of slope. Sub horizontal drainage is very effective in reducing pore-pressure along crack surfaces, or potential breakage surfaces. Vertical drainage is, generally, associated with sunken pumps which have the task of draining the water and lowering the ground water level. The use of continuous cycle pumps implies very high running costs, conditioning the use of this technique for only limited periods.

(iii) Chemical and mechanical methods - in which attempts are made to increase the shear strength of the unstable mass or to introduce active external forces (e.g., anchors, rock or ground nailing), or passive ones (e.g., structural wells, piles or reinforced ground) to contrast the destabilising forces. The most-used passive protection measures are: boulder-gathering trenches at the foot of the hillside, metal containment nets and boulder barriers. For defence against erosion, a series of solutions can be used, such as: geomats, geogrids and brushwood nets. These measures share the superficial character of their installation, given their low environmental impact. Geomats, or rather anti-eroding biomats or bionets, are purpose-made synthetic products for the protection and grassing of slopes, subject to surface wash, through two main erosion control mechanisms: the containment and reinforcement of the surficial ground; the protection from the impact of the raindrops. CRRI has established, through experiments carried out at different locations in the country, that surface erosion due to rainwater can be effectively controlled by promotion of vegetation on denudated hill slopes. For this purpose, use of natural geogrids made of coir has been found to be very successful and environment friendly. In this technique, the grass cover is established in one season and the natural fiber degrades in about 2-3 years. Asphalt mulch technique for promotion of vegetation has been used by CRRI. CRRI has used netlon geogrids to prevent rockfall and promote vegetation. Geogrid-Wicker, or brushwood mats, are made of vegetal material. Very long and flexible willow branches can be used, which are then covered with infill soil. Alternating stakes of different woody species are used and they are woven to form a barrier against the downward drag of the material eroded by the free water on the surface. Another category of work is the insertion of reinforcement elements in the ground such as: large diameterwells, anchors, network of micropiles, nailing, cellular faces or crib faces. Among the treatments belonging to the group of intrinsic improvements of a mechanical character in the ground, the technique of jet-grouting is often used. Another stabilisation technique is electro-osmotic treatment of the ground, but it is effective only in homogenous clayey grounds.

**Vulnerability Modification (Reducing the Impact)**

(i) **Forecasting and Warning-** For landslides it is possible to issue generalized warning of debris and mudflows following storms and heavy rainfall, but site-specific information is more difficult to obtain. Many landslides are preceded by a period of soil or surface creep before slope failure occurs, often giving rise to surface cracking. This process can be monitored with a view to provide a warning (Srivastava, 1998). The most common forms of monitoring include the use of inclinometers and tiltmeters to record evidence of increased hillslope activity, but this information is rarely formalized into official warning messages. There is need for instrumented monitoring of landslides, collection of relevant field data, and development of suitable methods to forecast the behavior of landslide, as well as to develop effective and economic corrective techniques. The standard method used to evaluate the potential of a landslide is the determination of the 'factor of safety'. A numerical value is determined for every factor related to the occurrence of a landslide. The factor of safety is a ratio in which all values that resist landslide are divided by the sum of all values that favor a landslide. The slope is considered stable when the factor of safety has a value that is greater than 1. Landslides are considered imminent when the value is less than 2. Myriad techniques and equipments are used to assess the instability of a slope. Conventional surveying methods measure and record the development of cracks, subsidence and uplift on slopes. Tiltmeters are used to record changes in the slope inclination near cracks and areas of weakness. Inclinometers and rock noise instruments are installed to record movements near cracks and ground deformations. Recording air temperature thresholds forecasts the onset of landslide brought on by snowmelts. Trends from past measurement, coupled with current monitoring of slope, increase the ability to predict future landslides. Monitoring rainfall and pore water pressure are other ways to predict potential landslides. The information of sub-surface geology by seismic refraction would help scholars to identify the nature of slope conditions, and to formulate future action plan, including preventive measures.

(ii) Hazard-resistant buildings- Slope stabilization, along with hazard resistant buildings, appears to be the most effective strategy. In this context, grading ordinances are an important tool. These normally require developers to obtain permits before they embark on earth-moving projects in hazard-prone areas. Ideally, they require reports from engineering geologists on proposed building sites before a local authority can approve plan. To work properly, this sort of system needs technically trained inspectors to enforce the regulations and a levy on development fees to become financially self-supporting. The vulnerability of buildings and infrastructure in landslide, however, is in most cases nearly 100 per cent, regardless of the quality of the construction. This option is, therefore, not highly relevant to landslide prone areas.

(iii) Land use planning -Landslide control is most effective when combined with land use planning. The recurrence of landslides at the same site means that land use zoning offers a practical method of hazard mitigation. Once potential sites have been identified, and frequency estimates made with initial mapping, law should explicitly encourage local communities to consider mass movement processes when undertaking land use changes. Relocation and regulation of functions is the most feasible option for landslide mitigation. It comprises a variety of measures, which are usually long term. Regulations may be applied to both existing and new or expanding settlements. A first and obvious step in the mitigation of landslide risks is to evaluate knowledge of landslides in the overall long-term planning procedures.

For the existing settlements, a more active approach may be required to reduce risks in the short term. In order to reduce risks, land use can be limited to functions that have lower risk. Limitation of permitted land use should be based on ranking of functions. This ranking of functions should be based on population density, function, and impact of damage and value of building/ infrastructure. Land use limitations may also aim at a reduction of the probability of landslide activation. These land use regulations are based on the fact that landslides are often activated by human actions. The construction weight, for example, may play an important role in the activation of landslides. The second type of land use limitations may, therefore, focus on limiting the building density and the construction weight (for example, by permitting light wooden constructions). When there is considerable risk of land sliding, the relocation of certain (or all) functions is the most feasible mitigation option. Avoiding hazardous sites is also, in practice, the most applied mitigation measure.

(iv) Community preparedness-People living in landslide prone areas need to note common warning signs of potential slope failure (Gupta and Verma, 2006). Some signs of landslides are doors or windows sticking or jamming for the first time. New cracks appearing in plaster, tile, brick, or the foundation of the house can be a precursor of earth movement. Widening cracks on paved streets or driveways also indicate movements in landslide areas. Sometimes underground utility lines will begin to break as a result of earth movement. Water will sometimes break through tho ground in new locations and fences, retaining walls, utility poles and trees will tilt more. A faint rumbling sound, increasing in volume, can be heard as the landslide nears. If any of these warning signs are experienced, evacuation plans should be made. It is recommended that there should be at least two planned evacuation routes, because roads may become inaccessible from deposit of slide materials. The community has to be trained to recognize and act upon the signals. A systematic training of the villagers in studying a combination of signs for evacuation could be the best preparedness for landslides. Contingency planning at the community level should include identification of safe shelters for evacuation of people and livestock. Identification of volunteers to work for search and rescue, and first aid training for treating the injured is required. Search and rescue teams should be trained in recovery techniques that are appropriate for landslides, along with rescue dogs.

Dissemination of technical, scientific and traditional knowledge may lessen the impact of disaster. Public awareness programs on causes and consequences of landslides should be organized. School textbooks should include chapter on landslides. A special newspaper section with emergency information on landslides should be published. Training for capacity building must be spread down to the grass root level for disaster management. The local government should be supported in efforts to develop and enforce land use and building ordinances in areas susceptible to landslides. Insurance coverage in the villages prone to landslides should be promoted. Awareness among policy makers at local level should be raised. Awareness among community leaders and general public, affected by landslide hazards, about the cost effectiveness and benefits of taking landslide hazard mitigation measures needs to be created too.

## CONCLUSION AND RECOMMENDATIONS

- Setting up of a State level database on landslides.
- Anticipatory planning based on carefully prepared landslide hazard assessment is required.
- Preparation of comprehensive landslide hazard zonation maps on specific scales (Singh, 2006).
- Prepare landslide susceptibility maps and landslide risk maps.
- Follow interdisciplinary approach to generate data on variables and use of remote sensing techniques and GIS.
- There is a need for a better translation of scientific information into user directed information. An appropriate approach would be the publication of both, an analytical map presenting observable features, and a synthetic map showing the interpretation in a simplified manner.
- Provide these maps to users.
- Update the maps.
- Impact assessment in holistic manner is required.
- Try event modification through geometric, hydrological, chemical, biological and mechanical methods.
- Through instrumentation and monitoring try to develop forecasting and warning systems.
- To reduce vulnerability focus on land use planning and community preparedness.

**REFERENCES**

Anbalagan, R. (1992). Landslide Hazard Evaluation and Zonation Mapping in Mountainous Terrain. *Engineering Geology*, 32: 269-277.

Anbalagan, R. (1996). An Overview of Landslide Hazards in Himalaya - Available Knowledge Base, Gaps and Recommendations for Future Research. *Himalayan Geology*, 17: 165-167.

Anbalagan, R. and Singh, B. (1996). Landslide Hazard Zonation Mapping- A Need for Sustainable Development of Uttaranchal with Special Reference to Route Locations. *Engineering Geology*, 32: 13-22.

Anbalagan, R., Singh, B., Chakarborty, D. and Kohli, A. (2007). A Field Manual for Landslide Investigations. Department of Science and Technology, New Delhi.

Gupta, 'V.' Sah, M.P., Virdi, N.S. and Bartarya, S.K. (1993). Landslide Hazard Zonation in the Upper Satluj Valley, District Kinnaur, Himachal Pradesh. *J. Himalayan Geology*, 4 (1): 81-93.

Gupta, S.S. and Verma, V.K. (2006). Landslide Hazards in Indian Hilly Areas, *In: Environmental Geo-Kazards- Science and Society*, (Eds. K.K. Sharma, S.K. Bandooni and V.S. Negi). Research India Press, New Delhi, pp. 162-170.

Rudra, K., Bandyopadhayay, G and Bandyopadyay, M.K. (1982). Landslides in Southern Sikkim. *In: Himalaya: Landforms and Processes* (Eds. V.K. Verma and P.S. Sakhlani). Today and Tomorrow's Publication, New Delhi, pp. 79-86.

Sarda, V.K. and Gupta, A.K. (2006). Geotechnical Investigations Related to Landslides, *In: Environmental Geo-hazards - Science and Society*, (Eds. K.K. Sharma, S.K. Bandooni and V.S. Negi). Research India Press, New Delhi, pp. 138-146.

Shrivastava, A.K. (1998). Landslide prevention and control measures in Sikkim, *In: Sikkim : Perspectives for Planning and Development*, (Eds. S.C. Rai, E. Sharma and R.C. Sundriyal). Sikkim Science Society, Gangtok, pp. 597-664.

Singh, B. (2006). Landslide Hazard Mitigation: DST's Initiatives, *In: Environmental Geo-hazards - Science and Society*, (Eds. K.K. Sharma, S.K. Bandooni and V.S. Negi). Research India Press, New Delhi, pp. 170-182.

Task Force Report on Environment and Development of Sikkim (1982). Ministry of Forests and Environment, Government of India, New Delhi.

Varnes, D.J. (1984). Landslide Hazard Zonation: A Review of Principles and Practices. *In: Natural Hazards: An IAEG Monograph.* UNESCO, Rome, pp. 1-55.

## Chapter 4

# Seismicity and Earthquake Prediction with Electromagnetic Waves

*Vinod Singh*[1], *A.K. Singh*[2], *and S.S. Verma*[3]

## INTRODUCTION

Earthquake is an intense shaking of the earth's crust induced by sudden release of strain, which results into the displacement of rocks along a fault beneath the earth's surface. Earthquake activity is also known as seismicity or seismic activity. It is one of the most sudden and destructive natural hazards, but unlike tornado and lightening its harmful effects are more extensive in nature. When some disturbance or displacement of rocks occurs at some depth below the surface, shock waves originate and travel in different directions causing vibrations. The point of origin of an earthquake below the earth's surface is called its *focus* or *hypocentre.* Now-a-days, focus is taken to be a zone rather than a point of origin (Bolt, 1993). The point on the earth surface that is the nearest or vertically above the focus is called as *epicentre.* It is the point where the vibrations of an earthquake reach first of all. Maximum damage to life and property is found to be concentrated here.

The origin or focus of an earthquake could be between a few hundred metres to several hundred kilometres below the surface. It has been observed that the depth of the focus or the distance from the epicentre to the focus, is often less than 50 kms, (called *shallow-focus earthquake),* very few earthquake originate between 300-700 km of depth *(deep-focus earthquake)* and no earthquake has been found to originate below 700

[1]Department of Geography, Govt. Dungar College, Bikaner (Raj.)-334001, India.

[2]Department of Physics, Dronacharya College of Engineering, Farukhnagar, Gurgaon (Haryana)-122001, India.

[3]Department of Physics, Sant Longowal Institute of Engineering & Technology, Longowal-1418106, Dist. Sangarur (Punjab), India.

km from the surface. The present article focuses light on intensity, magnitude and effects of earthquakes and their prediction with electromagnetic waves.

**Earthquake Waves**

Vibrations of the earthquake generally lead to short-lived ground movements which rarely last more than a minute. These elastic, seismic waves, generated at the focus, travel in different directions with their typical velocities. The route and pattern of these waves tells a lot about the interior structure of the earth too. Three main types of waves are P-waves, S-waves and the L- waves. The primary, *push and pull waves* or P-waves are the fastest waves which travel faster in the rigid rocks. These are also called longitudinal waves because the particles of the medium vibrate in the direction of the progress of the wave. The secondary or S-waves are also called the shear, distortional or transverse waves because the particles of the medium vibrate at right angles to the direction of propagation of the wave, just as in light waves. These waves cannot pass through liquid medium. The P and S waves are together referred to as *body-waves* because they travel deep into the body of the earth, undergoing refractions and reflections in the process, before re-emerging on the surface at a point far-off from the focus.

The L-waves or long waves are confined mainly to the surface of the earth. Hence, these waves are also known as *surface waves.* These surfaces waves are the slowest and are of two main types : in *Rayleigh waves* the displacement of the particles is partly in the direction of propagation and partly at right angles to it, while in *Love waves* the displacement of the particles of the medium is in direction of propagation of the waves. The Rayleigh waves tend to distort the surface into a wavy, zig-zag shape while the Love waves tend to create shearing or breaking ruptures (Singh, 2001). The L-waves are more hazardous to the structures than P and S waves.

**Causal Factors and Mechanisms**

Now-a-days, earthquakes' occurrence is explained in terms *of plate tectonics.* The earth's crust is divided into more than fifteen tectonic plates, which are rigid, 50 to 100 km thick, continental or oceanic masses in constant motion (Fig. 1). The plates are made of crust and upper mantle and may move at a rate of about 2 to 10 $cmyr^{-1}$. Different plate boundaries emerge due to relative motion of two plates : *Convergent boundaries* occur where two plates move towards each other and one plate is subducted under the other; *divergent boundaries* develop where two plates move away from each other, causing rift valleys and volcanic zones; *transformed boundaries* are found where two plates slide past each other laterally along fault lines. All the three types can lead to earthquakes but the convergent plate boundaries (circum-Pacific earthquake zone and Himalayan belt are the best examples) are especially responsible.

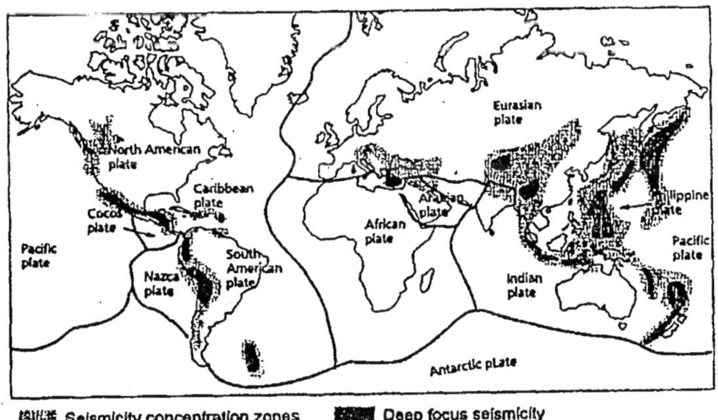


**Fig. 1: Global Seismicity Distribution: Margins of Crustal Plates are the Major Seismic Belts of the World.**

***Source:*** Park, 2001

Deep focused, intense earthquakes, occur along the convergent boundaries when a plate collides or goes down under the another. Divergent boundaries occur along the mid-oceanic ridges (MORs), where shallow depth seismicity, mostly of moderate to low intensity, occurs in association with vulcanism. Transformed boundaries occur along what Wilson (1965) called transform faults. Transform faults are very common near the mid-oceanic ridges. Along these mid-oceanic ridges, normal faults originate when two plates move apart in opposite directions. Transform faults are strike slip type faults which appear between the ridge segments, shifted perpendicularly to the axis of the MOR, due to the plate movements in opposite directions. San Andreas fault in California (USA) is a transform fault, though many now believe it to be the boundary found between the Pacific and the North American plates. Comparatively shallow-focused earthquakes occur along such faults or transformed boundaries (Strahler, 1981). Plate tectonics has been able to explain the earthquakes originating on plate margins, but has been unconvincing in its explanation of the intra-plate earthquakes.

Earthquakes may be caused by *non-tectonic* or *tectonic* reasons. Vibrations in the ground may be induced by non-tectonic factors like volcanic eruptions, underground nuclear explosions, water injection in the basal rock, and increase in crustal loading due to large water reservoirs related to dams. Vulcanism and underground nuclear explosions can land to minor seismic activity. The toxic liquid wastes from manufacture of chemical war products, were planned to be disposed off safely in a well, dug about 3800 m deep in the basal rocks near Denver city, in Rocky mountains of the USA. The wastewater injection was started in 1962, in a hitherto seismically insignificant region. Almost immediately, earthquakes began to rock the area. Investigation of the samples indicated the presence of deep underground faults. These long-dormant faults were seemingly reactivated due to lubricating action of the injected wastewater, resulting into earthquakes (Evans, 1966). Increased load on the earth's crust due to accumulated water in the reservoirs, associated with large river dams, can induce seismic activity. Reservoir associated seismicity is supposed to be created by the *increased pore pressure* in the basal rocks, which leads to the failure of the rocks, or to the reactivation of pre-existing faults in the rocks, causing slips and tremors. In highly susceptible areas the *sagging effect,* derived from the additional weight exerted by the newly established water reservoirs on the lower rocks, initiates a readjustment process signified by seismic activity. Reservoir associated seismicity has been reported from all over the world. Some famous examples are those related to Lake Mead on Colorado river (USA), lake Kariba (Zambia), Lake Nasser on the river Nile (Egypt) and Koyna Dam on Koyna river in Maharashtra (India). The last occurred on December 11,1967 on the shield area of the Peninsular Plateau of India. This plateau had, till then, been classified as stable and aseismic.

Most of the earthquakes are tectonic in nature, i.e., these are caused by the forces that lead to movements in the earth crust. When the strain becomes greater than the tolerance or bearing capacity of the rocks, faults or ruptures occur, along which slippage of massive blocks of the crust takes place, causing seismic movements. When the rocks are in such a delicate balance, even a slight disturbance can cause earthquakes. Such naturally or artificially generated earthquakes are called *triggered earthquakes.* The focus of the earthquake is a point below the earth surface where displacement of rocks takes place along the faults. Based on the pre and post earthquake observations of the San Francisco earthquake of 1906, H.F. Ried propounded the *elastic rebound theory* for tectonic earthquakes in 1911. Ried assumed the rocks to behave as elastic matter. Rocks bend in response to the stresses and, eventually, get ruptured when the maximum elastic limit is reached. The maximally displaced, and ruptured, block of rock rebounds elastically to its original position. During this process of sudden rebounding, most of the elastic energy accumulated due to the process of stress is released at the point of rock displacement (Fig. 2). The energy released from this point, or focus of the earthquake, is in the form of the earthquake or seismic waves (Singh, 2001).

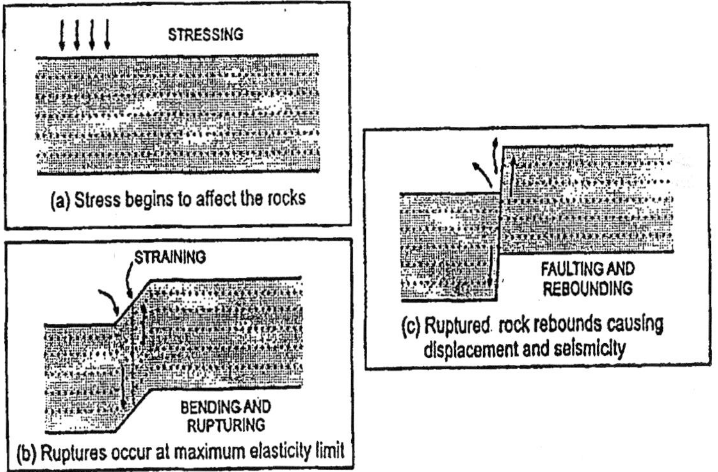


**Fig. 2. Elastic Rebound Theory of Earthquakes**

## Earthquake Hazards and Effects

The energy of an earthquake is measured by its magnitude, however, the hazard of an earthquake is related to its intensity. Magnitude is the severity of earthquake at its focus. It is calculated from the amplitude of the seismic waves and signifies the release of total energy at this point of origin. It is measured on a unit-less Richter scale. In this logarithmic scale, magnitude increase by one reflects a tenfold increase in the amplitude of the seismic waves. Compared to magnitude, intensity of an earthquake denotes the energy of seismic waves reaching a point of observation. Intensity decreases with distance from the focus and shows the impact on man and his structures (Table 1). Intensity is measured on various scales, one of which is the qualitative *modified Mercalli scale.* For a given earthquake, magnitude is fixed but intensity varies with distance (Strahler, 1981). Intensity can be represented on map by using iso-seismal lines, which are lines joining places of equal intensity of seismic waves around epicentre the zone of maximum intensity. Since destructive power of an earthquake is related to the type of underlying rocks, their structure as well as vicinity of the epicentre, isoseismal maps are useful in planning for mitigating the effects of earthquakes.

**Table 1: Intensity, Magnitude and Effects of Earthquakes**

| *Intensity (Mercalli number)* | *Mercalli name* | *Magnitude (Richter scale)* | *Effects* |
|---|---|---|---|
| I | Instrumental | <3.5 | Detected by seismograph only |
| II | Feeble | 3.5 | Noticeable by people at rest |
| III | Slight | 4.2 | Like vibrations from a moving truck |
| IV | Moderate | 4.5 | Normally felt indoors; rocks parked cars |
| V | Rather strong | 4.8 | Felt generally; most sleepers wake |
| VI | Strong | 5.4 | Trees swing; furniture moves |
| VII | Very strong | 6.1 | Walls crack; general alarm |
| VIII | Destructive | 6.5 | Walls fall; weak structures harmed |
| IX | Ruinous | 6.9 | Some houses collapse: ground cracks |
| X | Disastrous | 7.3 | Rails bend; many buildings ruined |
| XI | Very disastrous | 8.1 | Landslides; few buildings survive |
| XII | Catastrophic | >8.1 | Total ruin; wavy ground |

The hazards associated with earthquake can be grouped into *primary* and *secondary* hazards. Primary hazards occur alongwith the phenomenon of earthquake and include ground shaking, faulting and fracturing, and tectonic deformations. These result into building collapses, damage to the public utilities, changes in the underground water resources and courses of rivers. Secondary hazards occur at the end of the earthquake phenomenon and include, *inter alia*, land and mud slides, snow avalanches, soil liquefaction and tsunamis. Secondary hazards result into loss of property and life due to collapses, fires, diseases and floods. Buildings are damaged due to earthquakes. Although the foundations of the buildings move with the oscillating motion of the ground during the tremors, the upper portions tend to have delayed motion due to inertia. This delay leads to differential stresses, cracking and collapse. Liquefaction is a process in which water saturated sediments temporarily lose strength and start to behave like a viscous liquid. Liquefaction may occur during an earthquake if underlying rocks are shaken so violently that they start behaving like liquids (Park, 2001). The structures are damaged or tilted due to it because their foundations are weakened. *Tsunami* is a series of waves which originates due to a sudden disturbance of the sea floor by a massive earthquake, under water slide or volcanic eruption. Generally, seabed earthquakes trigger these waves, in which the sea water starts to rock vertically, causing heavy loss of life and property in the densely-settled coastal areas. The Indonesian seabed earthquake of December 26, 2004 (magnitude > 8.5) devastated the littoral areas of the Indian ocean. Such tsunami-causing seabed earthquakes occur where two oceanic plates collide and one of them subducts beneath the another (Doyle, 1995).

Often enough more damage to buildings is caused by the after shocks of the earthquake than the foreshocks and the main shock itself. This is explained by the fact that aftershocks tend to be more frequent and long lasting, although much weaker, than the main shocks. The amount of damage caused by seismic activity depends upon: (i) the concentration of population near the epicentre (ii) the force of the seismic wave (iii) the kind of the buildings people live in, and (iv) the density of the buildings. Combating the ill-effects of earthquake would require such measures as stopping construction of large buildings in high-risk zones through land-use planning, prediction and warning systems, introducing insurance schemes and ensuring safer building designs (Park, 2001). As the force exerted on a building due to seismic waves depends upon the movement of the ground as well as the weight of the building, it is safer to construct light buildings with lighter roofs (Srivastava, 1983).

## Earthquake Prediction

An important way of reducing damage from earthquakes and, thus, a significant component of disaster mitigation strategy, is prediction of seismicity hazard. Prediction of earthquakes has three components of earthquake location, timing and magnitude. The latter two are more difficult to predict than the former. Using inputs from geology, seismology and engineering, locational aspects of earthquakes are more easily predicted. Two major approaches to earthquake prediction are based on the analysis *of historical* trends and *precursory* phenomena. The former is based on the long-term behavioural pattern of the fault-lines, the most probable scenes of seismicity. The latter is based on the study of changes in physical phenomena (the precursors) before an earthquake (Brumbaugh, 1998).

### 1. *Long-term prediction*

Long-term earthquake prediction is based on the long duration, longitudinal study of the seismic activity of a region. Basing itself upon the periodicity of earthquakes and accumulation of strain in the rocks, long-term prediction attempts to specify the *earthquake risk* in a given area. Seismic risk is the probability of occurrence of a critical earthquake in the selected region during the projected period, and is determined from the seismic history of the concerned region. Since accurate forecasting of the timing of earthquakes is still not possible, the objective of long-term prediction is to identify regions where intensive observations can be made for medium and short-term prediction. Long-term prediction also enables us to take precautionary engineering measures like designing quake-resistant buildings and structures.

The earthquakes' occurrence has close relationship with the distribution of fault lines. Hence, faults identification and monitoring is a major task in prediction of earthquakes. Estimation of the maximum magnitude of earthquakes is made based upon the assessment of fault length. Using the *pattern recognition* technique it has been shown that strong earthquakes occur at the zones of intersection of major faults. Seismic hazard risk mapping can be helpful in the identification of areas with maximum risk. Intensity of the earthquake is shown on the maps by drawing isoseismal lines, which join the places of equal seismic intensity. Intensity depends on the magnitude of the earthquake, as well as its focal depth, distance from the epicentre and soil characteristics. The limitation of these maps is that they rely heavily on historical patterns, with the underlying assumption that the seismic activity will tend to recur in the same regions (Park, 2001). This method was proved inadequate by the occurrence of Koyna earthquake of 1967, in a tectonically "stable" zone of India. Hence, more emphasis needs to be given on the geological history of the concerned area, as well as identification of faults where movements have occurred recently (Srivastava, 1983).

It has been theorized that slow movement of rock masses along a fault zone causes micro-cracks due to strain. This cracking increases the volume of rocks in the process of *dilatancy.* Generally, ground water moves into the cracks of the dilatated rocks. This further increases the volume of the rockmass, a process which causes a dooming effect on the ground surface. The lifting effect of the ingested water tends to reduce the friction between the rock masses on the two sides of the fault triggering quick strike-slip motion, experienced at the focal point of the earthquake (Strahler, 1981). Deep-focus earthquakes are felt over a larger area than the shallow focused earthquakes, although the latter cause greater damage over a wider zone.

Thus, researchers are able to project a high probability of an earthquake in a certain area within about three decades. They can do this by : studying historical earthquake records, monitoring the motion of earth's crust by satellite, and measuring with strain monitors below the earth's surface. But this method just has not the case of short-term earthquake forecasting.

### 2. Medium and short-term prediction

Medium-term earthquake prediction is concerned with forecasting earthquakes a few months to one year in advance, while short-term prediction attempts to do this a few hours to days ahead of the seismic activity. Although the methodology is generally similar for the two, short-term prediction has relied on the observation and reporting of local, seismicity-related phenomena by the experts, and more so by the people. Hydrochemical changes in the properties of underground water may be seen before an earthquake. These include increase in the concentration of dissolved minerals 2 to 8 days ahead, change in the temperature, pressure and discharge of water. Unusual animal behavior as restlessness, fleeing from the area etc., has been reported since long. Cat fish could prove a reliable precursor of earthquakes (Rikitake, 1976). It has been observed that increase in the emission of radon gas from the wells near the active faults may take place just before an earthquake. Similarly, decrease in the electrical resistivity of rocks may occur during *dilatancy*, although its relationship with faulting is unclear. Several small earthquakes occurring in a short time are called *earthquake swarm.* Such swarm may precede a major earthquake, though these foreshocks are not invariable phenomenon and no major quake may occur following these (Strahler, 1981).

The subduction zones along the destructive plate boundaries are active seismic zones. In these zones, an area where seismic activity is lesser than its neighbourhood, is known as a *seismic gap.* Such gaps are the areas where strains are under the process of accumulation, and may, thus, cause a major earthquake in the future. Identification of seismic gaps, and their present status, has proved successful in prediction of seismicity. It can be said, however, that a combination of different observations can enable us to successfully predict seismicity hazards in the short-term (Table 2).

**Table 2: Combinatorial Observations for Short-Term Earthquake Prediction**

| *Instruments* | *Objectives of use* |
|---|---|
| Laser Survey | Measuring surface movement |
| Magnetometer | Recording of changes in the earth's magnetic field |
| Near-surface seismometer | Recording larger shocks |
| Satellite relay | To relay data instantaneously |
| Seismometer | Recording micro- seismicity |
| Sensors in wells | Monitoring changes in ground water levels |
| Strain meter | Monitoring surface deformation |
| Vibreosis truck | Creating shear waves to probe the earthquake zone |

**Source:** Parks, 2001

## Earthquake Prediction and Electromagnetic Waves

Accurate short-term forecasts are important to save lives and property and, at last, seem to be within reach. They will come not from the mechanical phenomena-measurements of the movements of the earth's crust but, rather, from electromagnetic phenomena. And, remarkably, these predictions will come from signals gathered not only at the earth's surface but also far above it, in the ionosphere. The sign of an impending quake is a disturbance in the extremely low frequency (ELF) (i.e., <300m Hz) and ultra low frequency (ULF) (< 1 Hz) radio band-noticed in the weeks and more dramatically in the hours before an earthquake. For decades, researchers have detected strange phenomena in the form of odd radio noise and eerie lights in the sky in the weeks, hours, and days preceding an earthquake. But only recently have experts started systematically monitoring those phenomena and correlating them to earthquakes. Both the lights and the radio waves appear to be electromagnetic disturbances that happen when crystalline rocks are deformed-or even broken-by the slow grinding of the earth in the process of an earthquake. Although a rock in its normal state is an insulator, this cracking creates tremendous electric currents in the ground, which travel to the surface and into the air. The details of how the current is generated still remains a mystery.

One theory is that the deformation of the rock destabilizes its atoms, freeing a flood of electrons from their atomic bonds, and creating positively charged electron deficiencies, or holes. Researchers have demonstrated through laboratory rock-crushing experiments that the breaking of oxygen-to-oxygen bonds in the minerals of a fracturing rock could produce holes. These holes manage to propagate through rock up towards the surface, while the electrons flow down into Earth's hot mantle. The movement of these charges causes changes in the rock's magnetic field and these changes propagate to the surface in the form of electromagnetic waves in ELF &ULF range.

Another theory is that the fracture of rock allows ionized groundwater thousand of metres below the surface to move into the cracks. The flow of this ionized water lowers the resistance of the rock, creating an efficient pathway for an electric current. The currents, thus, generated alter the earth's magnetic field around the earthquake zone.

The frequencies of these waves are so low-with wavelength of about 30,000 kilometre - that these can easily penetrate kilometre of solid rock and be detected at the surface. However, signals at frequencies above a few hertz, by contrast, would rapidly be attenuated by the ground and lost. Thus, there is a need to devise technologies to be able to detect and separate, using resonance technique, such low frequency signals before they are lost. These electromagnetic effects can be detected in a number of ways:

(i) Using ground-based sensors (a signal-axis search-coil magnetometer) to monitor changes in:

(a) the low frequency magnetic field, and

(b) the conductivity of air at the earth's surface as charge congregates on rock outcroppings and ionizes the air.

(ii) Researchers are planning to launch an experimental satellite (like Detection of Electro-Magnetic Emissions Transmitted from Earthquake Regions : DEMETER) designed to remotely monitor such magnetic changes and a larger, more sensitive satellite is in the design stages which will:

(a) monitor noise levels at ELF and ULF radio waves, and

(b) observe the infrared light that is emitted when the positive holes migrate to the surface and then recombine with electrons.

(iii) Even the existing Global Positioning System (GPS) may serve as part of an earthquake warning system. Sometimes the charged particles generated under the ground in the days and weeks before an earthquake, change the total electron content of the ionosphere-a region of the atmosphere above about 70 km, containing charged particles. If the ground is full of positively charged holes, it would attract electrons from the ionosphere, decreasing the airborne electron concentration over an area as much as 100 km in diameter and pulling the ionosphere closer to Earth. This change in electron content can be detected by alterations in the behavior of GPS navigation and other radio signals.

In this way, the connection between large earthquakes and electromagnetic phenomenon in the ground and in the ionosphere is becoming increasingly solid. Scientists around the world are looking at all of these phenomena and their potential to predict earthquakes accurately and reliably. Researchers in many countries, including China, France, Greece, Italy, Japan, Taiwan and the United States, are now contributing to the electromagnetic data by monitoring known earthquake zones. Using these phenomena for earthquake prediction will require a combination of satellite and ground-based sensors. This means that accurate earthquake detection might require a large number of magnetic field and air-conductivity sensors on the ground to integrate space and ground-based sensors to detect all these precursor signals-electronically detected ELF and ULF magnetic-field changes, ionospheric changes, infrared luminescence, and air-conductivity changes-along with traditional mechanism and GPS monitoring of movements of earth's crust.

When such a broad range of phenomena is monitored, spikes registered by different monitors detecting different types of signals would make forecasts more reliable. Forecasters may, then, be able to issue graduated warnings within weeks, days, and hours, declaring increasing threat levels as the evidence from different sensors begins to point in the same direction. The technical and financial problems related to earthquake prediction technology are expected to be worked out within the next few years, and then governments in active earthquake areas could install warning systems based on electromagnetic waves, saving lives and minimizing the chaos of earthquakes. Alongwith electromagnetic seismicity indicators, however, other long-term and short-term earthquake precursors like thermal, water, animal and human will continue to hold and play their significant roles.

## CONCLUSION

Earthquake or seismic activity is the intense shaking of earth's crust due to release of accumulated strain in the rocks below the earth's surface. This strain results into rupture and displacement of rocks, generating elastic shock waves which radiate outwards from this point of origin, or focus. Of these, surface or L-waves originate when the rapidly-moving P-waves and the slower S-waves reach the surface of the earth. Heavy damage to the structures is caused by the slow up and down motion of these L-waves. Seismicity can be explained in terms of plate tectonics. Plates are the mobile units of earth comprising the crust and upper mantle. Converging plate boundaries give rise to deep-focus earthquakes, while the divergent and transformed plate boundaries lead to shallow-focus earthquakes. Normally, the shallow focus earthquakes are more damaging due to their proximity to the surface. However, plate tectonics fails to explain intra-plate seismicity. Earthquake can be triggered by natural forces as well as human activities. Included in the latter are water injection and reservoir-induced seismicity. Elastic rebound theory has been a good general theory, to explain mechanism of tectonic earthquakes, i.e., earthquakes caused by movements of rock-masses.

Seismicity hazard is a function of its *intensity*, which varies with distance from the *epicentre*, the place vertically above the focus. Intensity can be represented by *isoseismals*, the lines joining places of equal intensity around an epicentre. Hazards of an earthquake can be primary or secondary. The latter originate after the seismic phenomenon has occurred and include tsunamis, slides and avalanches. Earthquake prediction is, thus, one of the significant component of disaster management for mitigating the destruction to life and property. The long-term prediction bases itself upon the study of historical trends of seismicity, upon the accumulation of strain in the rocks, and the pattern of fault distribution. Medium, and especially short-term prediction is more important to save life and property. Short-term prediction is based on the observation of local, seismicity-related precursors like hydrochemical changes, unusual animal behavior, radon gas emission and electrical resistivity changes, etc. A combination of various observations can enable a more successful prediction. Now-a-days, short-term prediction based on electromagnetic waves is of prime concern. Scientists are investigating phenomena like disturbances in radio waves and eerie lights in the sky that precede hours to weeks before earthquakes. These phenomena are, now, in the process of being integrated into short-term prediction and, hopefully, will provide a firm basis for successful seismic prediction and disaster reduction in the days to come.

## REFERENCES

Bolt, B.A. (1993). *Earthquakes : A Primer.* W.H. Freeman, Oxford.

Brumbaugh, D.S. (1998). *Earthquakes : Science and Society.* Prentice Hall, Hemel Hempstead.

Doyle, H. (1995). *Seismology.* John Wiley & Sons, London, UK.

Evans, D.M. (1966). Man made Earthquakes in Denver. *Geotimes*, May-June, pp. 11-18.

http://www.nature.eom/nature/debates/earthquake/equake.7.html

http://origin.www. spectrum.ieee.org/login

http://www.spectrum.ieee.org/print/3275

Park, Chris (2001). *Environment : Principles and Applications.* 2nd edn., Routledge, London, pp. 160-173.

Rikitake, T.C. (1916). *Earthquake Prediction.* Elesevier Scientific Publishing Co., Amsterdam.

Singh, Parbin (2001). *Engineering & General Geology.* 6th edn., S.K. Kataria & Sons, Delhi, pp. 537-564.

Srivastava, H.N. (1983). *Forecasting Earthquakes.* National Book Trust, India, New Delhi, pp. 1-147.

Strahler, Arthur N. (1981) *Physical Geology.* Harper & Row Publishers, New York, pp. 183-305.

Wilson, J.T. (1965). A new class and their bearing on continental drift. *Nature*, 207.

# Chapter 5

# Floods and their Management with Special Reference to India

*S.K. MATHUR*[1], *K.K. SINGH*[2], *ASHA SHARMA*[3] *and IRA CHOWDHARY*[4]

## INTRODUCTION

Floods are one of the major natural disasters in India causing loss of life and property. These occur almost every year in many parts of the country. Flood is defined as a rising and overflowing of a water body onto a normally dryland. It is an overflow of an expanse of water that submerges land which is normally not submerged. It usually occurs when volume of water exceeds in a water body, such as river or lake, and results in water flow. The occurrence of flood usually takes place in the aftermath of meteorological events viz ; (i) intensive and prolonged rainfall (ii) unusual high coastal and estuarine water due to storm surges, seisches, etc. Seismic activities are also responsible for floods as well as Tsunamis that cause floods like a monster spreading across the sea as faster jet aeroplane. Normally, an intense earthquake starts Tsunami waves, rather series of waves, with a speed of more than 800 km $hr^{-1}$ crossing vast ocean. Their height is, sometimes, more than 30 m, thereby causing floods, inundating large coastal areas. These floods cause serious damages to growing crops, agricultural lands, settlements/habitation, roads, bridges, flood

[1] Quality Control Laboratory, Krishi Bhawan, Bikaner (Raj.) - 334002, India

[2] Project Directorate (Research), Agriculture & Soil Survey, Krishi Bhawan, Bikaner (Raj) - 334003, India.

[3] Lady Elgin Govt. Girls' Senior Secondary School, Bikaner (Raj.) - 334005, India.

[4] Sophia Sr. Secondary School, Bikaner (Raj.) - 334001, India.

control structures, irrigation dikes, etc. The flood prone area in India in India is estimated to be 40 million ha. Out of the total area liable to floods, nearly 32 million ha could be provided with reasonable protection, and approximately 50 per cent of such area has been provided with a reasonable protection so far through various means of flood management measures (Chandra 2003). Shukla (2007) had studied the floods of Mumbai (July, 2005) and disastrous floods of Bihar. It was observed that Mumbai floods rendered hundreds dead and thousands uprooted and made homeless in Bihar. There was mass destruction due to silting and overflowing of rivers leading to inundation of river catchments.

The article presents an overview of flood affected areas in India and damages caused to crops, human lives and public utilities. In addition, various approaches to management of floods are also dealt with.

## CAUSES OF FLOODS

The floods are often caused by :

- Rapid melting of snow and ice in the mountain areas,
- Cloudburst / heavy downpour in the mountain areas,
- Glacial lake outburst in the high Himalaya, and
- Failure of landslide/debris flow caused dams in high and rugged mountain areas.

Large mountain floods are characterized by enormous energy and high flow velocity combined with substantial bed load and debris flow, sudden rise and rapid decline in water level and severe erosion and deposition processes leading to catastrophic consequences.

## ESTIMATES OF FLOOD DAMAGES IN INDIA

Flood affected areas and flood damages to crops in India are presented in Table 1, and Figs. 1 and 2. The data reveals that damage to crops was of value Rs. 4,246.62 crore in the year 2000 followed by Rs. 755.67 crore in the year 2004. The houses damaged were 35,08,000 in number, followed by 1,54,4000 in years 1978 and 2004, respectively. The maximum damage to houses was of value Rs. 1307.89 crore in 1995. The area affected was 17.50 million ha during 1978 and 8.47 million ha during 2004. The human lives lost were 11,316 during 1977 and 1650 during 2004. The cattle lost were 6,18,000 during 1979 and 67,000 during 2004. The maximum population affected was 70.45 million in the year 1978 and maximum damage to crops was in the area of 10.15 million ha in 1988, crop damage was of value Rs. 4,246.62 crore during 2000 and the cattle lost was 6,18,000 in number during 1979 followed by 67,000 during 2004. The total damage to crops, houses and public utilities was recorded highest during the year 2000, that was of value Rs. 8,864.54 crore, followed by Rs. 3,854.65 crore during 2004.

The area affected due to floods was 16.30 million ha during 1988 followed by 11.80, 10.80 and 10.70 million ha in the years 1993, 1998 and 1984, respectively. The maximum area affected by floods was in the year 1988 and minimum during 1992 (Fig. 1). Damage to crops and total damages were recorded maximum in the year 1971 at 1993-94 prices, while maximum damage to crops and total damages were observed in the year 1961 (Fig. 2).

## FLOOD AFFECTED AREAS IN INDIA

The 1934 earthquake in North Bihar and the 1950 earthquake in Assam created obstructions in the rivers by land slip which led to terrible flooding. Almost all the coastal areas and islands are susceptible to heavy flooding. Apart from North-eastern part of our country, Tamil Nadu, Andhra Pradesh, Rajasthan, Gujarat have also experienced floods. However, the nature and magnitude of flood varies from years to year and place to place. It has been estimated that 40 million ha of landmass is flood prone in our country and, on an average, 8 million ha is affected by floods every year (De *et al.*, 2005). Shukla (2007) has observed that in Himalayan rivers, besides the heavy rainfall, the floods are often due to the formation of barriers in their course. These may be due to landslip, accumulation of vegetation and glacial moraines. It has been reported by Niwas and Khichar (2010) that flash flood can occur after a period of drought when heavy rain falls on dry and hard soils, and rainfall exceeds the infiltration rate and cause flooding alongwith slope. Damages caused by floods in different states are presented in Table 2. The maximum damage caused by floods was in Bihar (23.9%) followed by Uttar Pradesh (23.8%), Arunachal Pradesh (15.4%), West Bengal (7.0%), Gujarat (5.7%), Orissa (4.5%). The minimum percentage of damage was observed in Madhya Pradesh (1.3%) (Thakur, 2008).

Table 1: Flood Damages in India from 1953 to 2004

| S. N. | Years | Area Affected (Million ha) | Population Affected (Million) | Damage to Crops | | Cattle Lost Nos. (*000) | Damage to Houses | | Human Lives Lost (No.) | Damage to Public Utilities (Rs. Crore) | Total Damages to Crops, Houses & Public Utilities (Rs. Crore) |
|---|---|---|---|---|---|---|---|---|---|---|---|
| | | | | Area (Million ha) | Value (Rs. Crore) | | Nos (*000) | Value (Rs. Crore) | | | |
| 1 | 1953 | 2.29 | 24.28 | 0.93 | 42.08 | 47 | 265 | 7.42 | 37 | 2.90 | 52.40 |
| 2 | 1954 | 7.49 | 12.92 | 2.61 | 40.52 | 23 | 200 | 6.56 | 279 | 10.15 | 57.23 |
| 3 | 1955 | 9.44 | 25.27 | 5.31 | 77.80 | 72 | 1667 | 20.95 | 865 | 3.98 | 102.73 |
| 4 | 1956 | 9.24 | 14.57 | 1.11 | 44.44 | 16 | 726 | 8.05 | 462 | 1.14 | 53.63 |
| 5 | 1957 | 4.66 | 6.76 | 0.45 | 14.12 | 7 | 318 | 4.98 | 352 | 4.27 | 23.37 |
| 6 | 1958 | 6.26 | 10.98 | 1.40 | 38.28 | 18 | 382 | 3.90 | 389 | 1.79 | 43.97 |
| 7 | 1959 | 5.77 | 14.52 | 1.54 | 56.76 | 73 | 649 | 9.42 | 619 | 20.02 | 86.20 |
| 8 | 1960 | 7.53 | 8.35 | 2.27 | 42.55 | 14 | 610 | 14.31 | 510 | 6.31 | 63.17 |
| 9 | 1961 | 6.56 | 9.26 | 1.97 | 24.04 | 16 | 533 | 0.89 | 1374 | 6.44 | 31.37 |
| 10 | 1962 | 6.12 | 15.46 | 3.39 | 83.18 | 38 | 514 | 10.66 | 348 | 1.05 | 94.89 |
| 11 | 1963 | 3.49 | 10.93 | 2.05 | 30.17 | 5 | 421 | 3.70 | 432 | 2.74 | 36.61 |
| 12 | 1964 | 4.90 | 13.78 | 2.49 | 56.87 | 5 | 256 | 4.59 | 690 | 5.15 | 66.61 |
| 13 | 1965 | 1.46 | 3.61 | 0.27 | 5.87 | 7 | 113 | 0.20 | 79 | 1.07 | 7.14 |

Table 1 contd.

| 14 | 1966 | 4.74 | 14.40 | 2.16 | 80.15 | 9 | 217 | 2.54 | 180 | 5.74 | 88.43 |
|---|---|---|---|---|---|---|---|---|---|---|---|
| 15 | 1967 | 7.12 | 20.46 | 3.27 | 133.31 | 6 | 568 | 14.26 | 355 | 7.86 | 155.43 |
| 16 | 1968 | 7.15 | 21.17 | 2.62 | 144.61 | 130 | 683 | 41.11 | 3497 | 25.37 | 211.10 |
| 17 | 1969 | 6.20 | 33.22 | 2.91 | 281.90 | 270 | 1269 | 54.42 | 1408 | 68.11 | 404.44 |
| 18 | 1970 | 8.46 | 31.83 | 4.91 | 162.78 | 19 | 1434 | 48.61 | 1076 | 76.44 | 287.83 |
| 19 | 1971 | 13.25 | 59.74 | 6.24 | 423.13 | 13 | 2428 | 80.24 | 994 | 129.11 | 632.48 |
| 20 | 1972 | 4.10 | 26.69 | 2.45 | 98.56 | 58 | 897 | 12.46 | 544 | 47.17 | 158.19 |
| 21 | 1973 | 11.79 | 64.08 | 3.73 | 428.03 | 261 | 870 | 52.48 | 1349 | 88.49 | 569.00 |
| 22 | 1974 | 6.70 | 29.45 | 3.33 | 411.64 | 17 | 747 | 7243 | 387 | 84.94 | 569.02 |
| 23 | 1975 | 6.17 | 31.36 | 3.85 | 271.64 | 17 | 804 | 34.1 | 686 | 166.05 | 471.64 |
| 24 | 1976 | 11.91 | 50.46 | 6.04 | 595.03 | 80 | 1746 | 92.16 | 1373 | 201.50 | 888.69 |
| 25 | 1977 | 11.46 | 49.43 | 6.84 | 720.61 | 556 | 1662 | 152.29 | 11316 | 328.95 | 1201.85 |
| 26 | 1978 | 17.50 | 70.45 | 9.96 | 911.09 | 239 | 3508 | 167.57 | 3396 | 376.10 | 1454.76 |
| 27 | 1979 | 3.99 | 19.52 | 2.17 | 169.97 | 618 | 1329 | 210.61 | 3637 | 233.63 | 614.20 |
| 28 | 1980 | 11.46 | 54.12 | 5.55 | 366.37 | 59 | 2533 | 170.85 | 1913 | 303.28 | 840.50 |
| 29 | 1981 | 6.12 | 32.49 | 3.27 | 524.56 | 82 | 913 | 159.63 | 1376 | 512.31 | 1196.50 |
| 30 | 1982 | 8.87 | 56.01 | 5 | 589.40 | 247 | 2397 | 383.87 | 1573 | 671.61 | 1644.88 |
| 31 | 1983 | 9.02 | 61.03 | 3.29 | 1285.85 | 153 | 2394 | 332.33 | 2378 | 873.43 | 2491.61 |

*Table 1 contd.*

| | | | | | | | | | | | |
|---|---|---|---|---|---|---|---|---|---|---|---|
| 32 | 1984 | 10.71 | 54.55 | 5.19 | 906.09 | 141 | 1764 | 181.31 | 1661 | 818.16 | 1905.56 |
| 33 | 1985 | 8.38 | 59.59 | 4.65 | 1425.37 | 43 | 2450 | 583.86 | 1804 | 2050.04 | 4059.27 |
| 34 | 1986 | 8.81 | 55.50 | 4.58 | 1231.64 | 60 | 2049 | 534.41 | 1200 | 1982.54 | 3748.53 |
| 35 | 1987 | 8.89 | 48.34 | 4.94 | 1154.62 | 129 | 2919 | 464.49 | 1835 | 950.59 | 2569.72 |
| 36 | 1988 | 16.29 | 59.55 | 10.15 | 2510.90 | 151 | 2277 | 741.60 | 4252 | 1377.80 | 4630.30 |
| 37 | 1989 | 8.06 | 34.15 | 3.01 | 956.74 | 75 | 782 | 149.82 | 1718 | 1298.77 | 2405.33 |
| 38 | 1990 | 9.30 | 40.26 | 3.18 | 695.61 | 134 | 1020 | 213.73 | 1855 | 455.27 | 1708.92 |
| 39 | 1991 | 6.36 | 33.89 | 2.70 | 579.02 | 41 | 1134 | 180.42 | 1187 | 728.89 | 1488.33 |
| 40 | 1992 | 2.64 | 19.26 | 1.75 | 1027.58 | 79 | 687 | 308.28 | 1533 | 2010.67 | 3344.53 |
| 41 | 1993 | 11.44 | 30.41 | 3.21 | 1308.63 | 211 | 1926 | 528.32 | 2864 | 1445.53 | 3382.49 |
| 42 | 1994 | 4.81 | 27.55 | 3.96 | 888.62 | 52 | 915 | 165.21 | 2078 | 740.76 | 1794.59 |
| 43 | 1995 | 5.24 | 35.93 | 3.24 | 1714.87 | 62 | 2002 | 1307.9 | 1814 | 679.63 | 3702.31 |
| 44 | 1996 | 8.05 | 44.73 | 3.83 | 1124.49 | 73 | 727 | 176.59 | 1803 | 861.39 | 3005.74 |
| 45 | 1997 | 4.57 | 29.66 | 2.26 | 692.74 | 28 | 505 | 152.50 | 1402 | 1985.93 | 2831.18 |
| 46 | 1998 | 10.85 | 47.44 | 7.45 | 2594.21 | 107 | 1933 | 118.80 | 2889 | 5157.77 | 8860.72 |
| 47 | 1999 | 7.77 | 27.99 | 1.75 | 1850.87 | 91 | 1613 | 1299.10 | 745 | 462.83 | 3612.76 |
| 48 | 2000 | 5.38 | 45.01 | 3.58 | 4246.62 | 123 | 2629 | 680.94 | 2606 | 3936.98 | 8864.54 |
| 49 | 2001 | 6.18 | 26.46 | 3.96 | 688.48 | 33 | 716 | 816.47 | 1444 | 5604.46 | 7109.41 |

*Table 1 contd.*

| | | | | | | | | | | | |
|---|---|---|---|---|---|---|---|---|---|---|---|
| 50 | 2002 | 7.09 | 26.32 | 2.19 | 913.09 | 22 | 762 | 599.37 | 1001 | 1062.08 | 2574.54 |
| 51 | 2003* | 6.5 | 34.47 | 3.43 | 1424.83 | 16 | 847 | 802.93 | 1864 | 2206.60 | 4434.35 |
| 52 | 2004 | 8.47 | 34.19 | 2.92 | 755.67 | 67 | 1544 | 870.73 | 1650 | 2228.25 | 3854.65 |
| Total | | 397 | 1712 | 185 | 36846 | 4913 | 64252 | 13054 | 83079 | 42312.00 | 94488 |
| Average | | 7.63 | 32.92 | 3.56 | 708.57 | 94 | 1235.61 | 251.05 | 1597.67 | 813.69 | 1817.07 |
| Maximum (Year) | | 17.50 (1978) | 70.45 (1978) | 10.15 (1988) | 4246.62 (2000) | 618 (1979) | 3508 (1978) | 1307.89 (1995) | 11316 (1977) | 5604 (2001) | 8864.54 (2000) |

Source : Directorate of Flood Management and Planning, Central Water Commission, New Delhi.

* Figures are tentative.

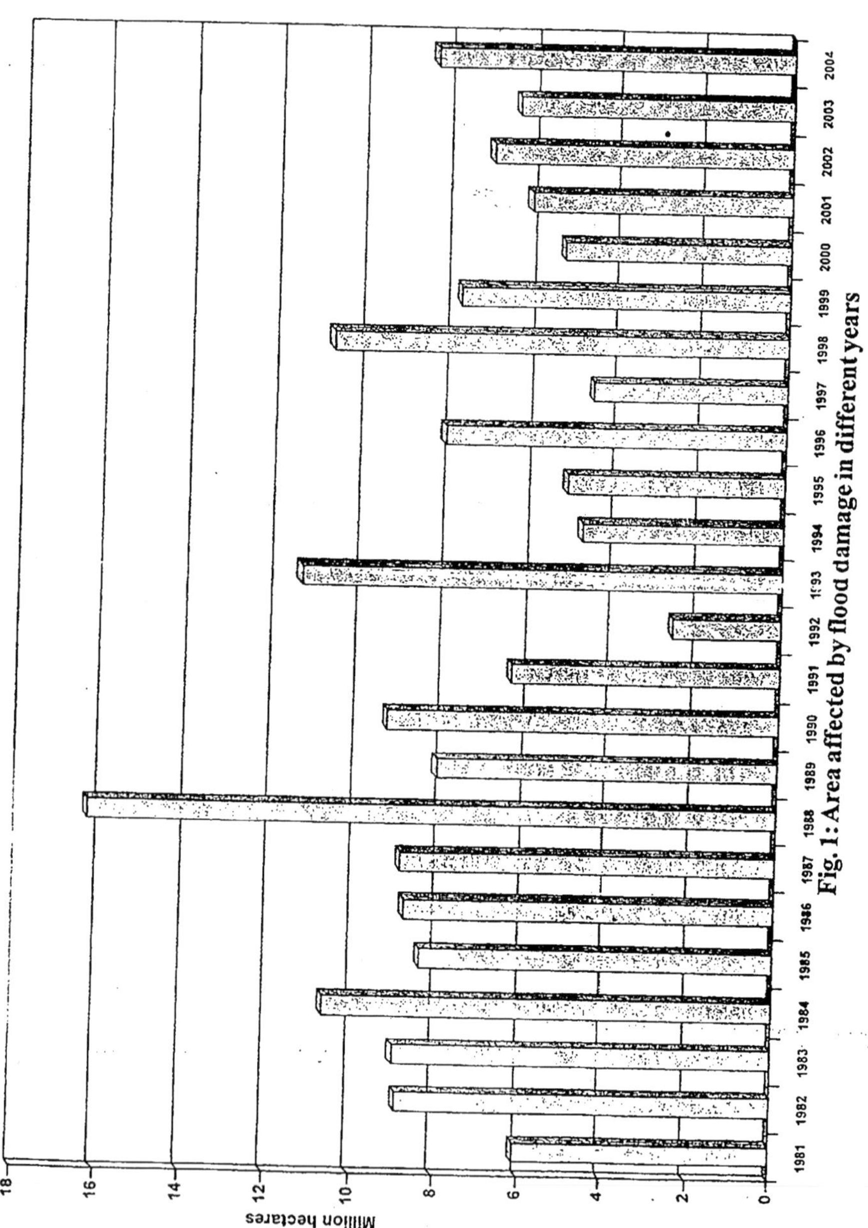


**Fig. 1: Area affected by flood damage in different years**

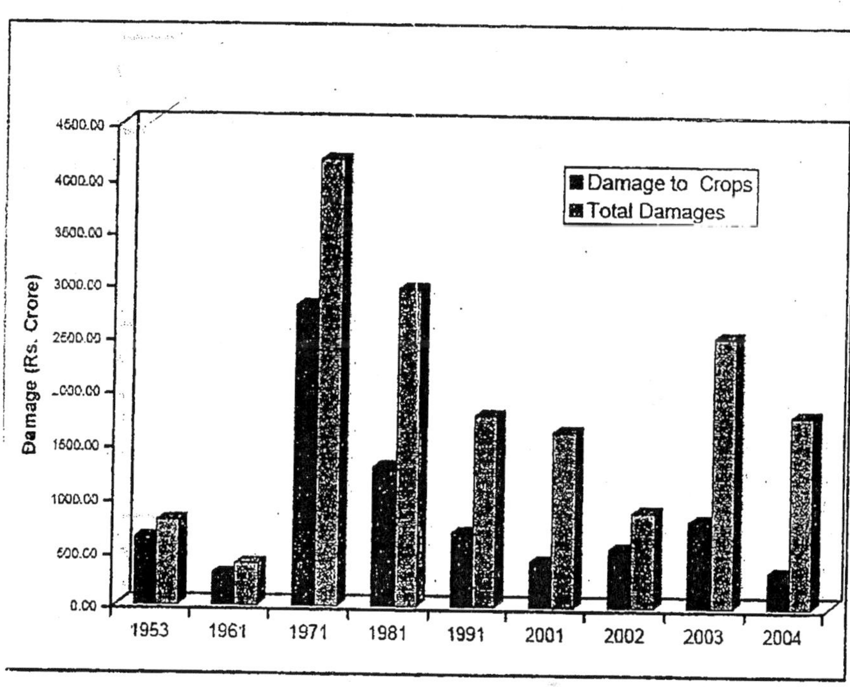


**Fig. 2: Trends in flood damages, 1981-2004**

**Table 2: State wise average damages caused by floods (1971-78)**

| *State* | *Damages(%)* | *State* | *Damages(%)* |
|---|---|---|---|
| Bihar | 23.9 | Rajasthan | 4.5 |
| Uttar Pradesh | 23.8 | Tamil Nadu | 3.8 |
| Arunachal Pradesh | 15.4 | Haryana | 3.2 |
| West Bengal | 7.0 | Assam | 2.1 |
| Gujarat | 5.7 | Punjab | 1.4 |
| Orissa | 4.5 | Madhya Pradesh | 1.3 |

**Source :** Thakur, 2008

Thus, in recent years, there is a significant increase in the frequency and intensity of floods in India because of rapid rate of deforestation in the catchments of major rivers and their tributaries, and consequent accelerated rate of soil erosion, increase in sediment load of rivers, silting and rise of river beds and marked reduction in water accommodating capacity of river valleys; encroachment on agricultural lands on valley sides, even down to their channels, etc. (Thakur, 2008). Major flood affected areas in India are discussed below :

**(i) Floods in North-eastern region**

The Brahmaputra is one of the largest rivers of our country. The origin of this river is in East of Mansarovar and the river is known as Tsangpo in Tibet. It enters in India near Tuting in Arunachal Pradesh after flowing 1625 km through Tibet. After flowing 918 km in Arunachal Pradesh and Assam, it flows 258 km in Bangladesh and then joins the Ganges river prior to its outfall in the sea. During this long run, it meets with tributaries like Subansiri, Kameng, Dhansiri, Manas, Champamati, Saukosh, Noa Dihing, Buri Dihing, Dibang, Dikhu and Kopili. Floods are regular annual feature in Brahamaputra basin. Every year it is observed that vast area is inundated by severe floods for at least three months (Ghani, 2003; Shukla, 2007; Bhuyan, 2007).

**(ii) Floods in North-western region**

Indus is the most important river of this region, however, most of the area of Indus river is now in Pakistan, but upper courses of Indus are in India which cover Jammu & Kashmir. In this region, the floods are also caused by the Jhelum, the Chenab, the Ravi and the Beas, which results in submergence and water logging of vast stretches. The floods in the year 2000 were devastating in Himachal Pradesh. Millions of people became homeless, roads were destroyed leading to disruption in supply of basic amenities to the huge population of affected area (Shukla, 2007).

**(iii) Floods in Western India**

The flood prone areas in Rajasthan are basins of the river Luni in Ajmer, Barmer, Jalore, Jodhpur and Pali districts, the Sukdi in Sirohi, the Banas in Sirohi, Udaipur, Chittorgarh, Bundi and Jaipur districts, the Mahi in Banswara, the Chambal in Kota, Jhalawar and Baran districts, the Banganga in Bharatpur, the Sabi in Alwar, the Shekhawadati in Nagaur and the Ghaggar in Hanumangarh and Sriganganagar districts. Floods have been highly destructive even in the semi-arid and arid regions of Rajasthan. Rajasthan experienced very heavy rainfall of 820 mm in ten days and about

10 districts were severely affected. In July 2003, heavy rainfall in Sikar district also resulted in devastating floods (Shukla 2007). Barmer district in Rajasthan experienced heavy rainfall (400 mm) during 20[th] to 23[rd] August, 2006. As such, all the water bodies got filled and epithermal streams drained heavy run -off to the low lying areas. All the inter-dunal flats and depressions got submerged resulting in heavy damages to human, livestock, property and infrastructure particularly in Kawas / Sar Ka Par and Malwa. The situation did not improve even after 20 days, as the standing water could not percolate due to sub surface impervious layer of clay and gypsum (Sharma *et al.* 2006).

**(iv) The Peninsular Region**

The floods, on the other hand, are not frequent in Southern part of the country as compared to Northern region. There are four major rivers viz., Krishna, Kaveri, Mahanadi and Godavari. All these rivers flow Eastward and fall into the Bay of Bengal. Sometimes due to heavy rains in Western Ghats, these rivers cause floods in the area. Almost all the rivers of Northern India receive water from melting of ice as well as from heavy rains whereas the rivers of South receive water only from rains. These rivers have gentle slopes and, by and large, have straight and linear courses. As such, the floods caused by them are not so furious as those caused by Northern rivers. However, flooding in coastal regions is a regular feature because of sea waves. Cyclones, heavy storms and tidal waves are common features of seas which are the main reasons for floods in coastal areas (Shukla, 2008).

Tsunami floods of 2004 were devastating. These floods were a result of a very severe earthquake that occurred off Sumatra, an island of Indonesia. India was badly hit by these waves and more than 11,000 people from coastal areas lost their lives. The huge flood was so fast that everybody was taken aback in panic. People could not understand and take stock of the situation. (Verma, 2008).

Flood damages are regular feature in Mumbai. It may be noted that flood of the year 2005 was severely disastrous. In 2005 heavy rains covered entire Maharashtra state and various parts of Karnataka state. However, Mumbai, Thane and Kalyan were affected badly (Kulshrestha, 2000).

From the ongoing discussion, it has been observed that Northern plains and North-eastern areas of our country are highly vulnerable to floods, as compared to peninsular India.

**(v) Floods in Uttar Pradesh**

The severely flood affected districts of Uttar Pradesh are Lucknow, Allahabad, Jaunpur and Gorakhpur. It has been experienced that Lucknow is prone to severe flood due to over flow in Gomti River which has 8826.68 km $^2$ of catchment area. The flood water of Gomti River enters in the low lying areas on both sides of the river. The intensity of flood varies form year to year whereas highest gauge recorded was 113.2 m and corresponding discharge was 150,000 cusec in 1960 (Thakur, 2008). Almost similar flood situations were experienced in 1971 and 1980.

Similarly, Allahabad is also adversely affected by floods due to its situation on the confluence of two mighty rivers the Ganges and the Yamuna. Thakur (2008) reported that Ganges and Yamuna water crossed danger mark (84.75 m) five times within a span of 13 years (1971-1983) at Allahabad.

Flood damages are also perpetual in Jaunpur district of Uttar Pradesh. The records of floods in Jaunpur districts indicated high magnitude of flood havoc particularly in 1955, 1960, 1971 and 1980. The danger level has been fixed at 74.66 m at gauge station one km upstream of Shashi bridge. The data indicated that level of 1980 flood (77.25 m) was though lower than 1971 flood (77.73 m), yet it surpassed all the previous records of duration, dimension and horizontal extent of the flood water as 2/3$^{rd}$ part of the city reeled under water for about 60 days on stretch, when Gomti river crossed the danger level on July 2, 1980 and a sign of relief was seen only after September 10, 1980 but the clearance of water took another 15 days (Thakur, 2008).

Gorakhpur district is situated in North-eastern corner of the state. The geography of the district is unique with various types of terrains and high susceptibility to floods. In the North, the Ghaghra River flows where as the Rapti river flows on Western side of the city. Heavy rainfall is the main cause of devastating floods in Gorakhpur district. The records of floods indicated that the magnitude was very high in the year 2000 and the assistance of Army and Air force was called by civil authorities for rescue and evacuation of the people. More than 4 million people became homeless suddenly. However, government along with Army, Air Force and voluntary organizations did commendable task towards relief and rescue operations. Medical facilities were also rushed to the affected area.

**(vi) Floods in Bihar**

The devastation caused by Kosi river that is also called "Sorrow of Bihar" in the month of August 2008, ruined more than 2.5 millions people and their houses in 15 districts, including a number of townships, 242 panchayats and 671 villages which lie along the new course of the river. The Prime Minister Dr. Man Mohan Singh has declared it a National Calamity (Anonymous, 2008). Sinha (2008) has reported that this calamity has cast shadow on India's ties with Nepal since the genesis of the calamity could well lie in Katmandu's apathy towards timely maintenance of barrage that was breached. The gap has widened to 1.7 km and it is already 3.0 km wide. As such, bridging it would require a mammoth technical operation. Of the total 0.144 million cusec flow of Kosi river through barrage the river, after the breach, was out pouring a massive 0.118 million cusec through its new route crossing through Bihar plains.

In Uttar Pradesh and Bihar almost every year the mighty Ganges receives a large number of tributaries such as the Gomti, the Ghaghra, the Gandak and the Kosi from left and from the right the Yamuna and the Sone rivers. This brings large quantities of water resulting in devastating floods (Shukla, 2008).

## MANAGEMENT OF FLOODS

Since time immemorial, civilizations have prospered on flood plains, deriving benefits from floods. At the same time, floods have resulted in untold miseries for millions of people around the world. Despite enormous environmental and economic benefits, floods continue to be the leading natural disaster causing loss of life and affecting sustainable development. Flood management measures in the past have largely been successful in mitigating the adverse impacts of floods but have often created disparities owing to inappropriate policies and inadequate attention paid to social issues. In many countries, in certain areas that are frequently flooded, "living with floods" has been a major strategy. However, the benefits of overall economic development in other parts of these countries have yet to reach such frequently flooded areas (Naithani, 2003).

It is possible to live safely in such places, if the following rules are followed :

1. Keep enough space on both sides of the torrent, since the devastating forces in the centre are too strong to be managed.
2. Try to dissipate the energy by distributing the flowing water over a large area. By doing this, the flooding depth lowers and the water can be guided by small, inexpensive structures.
3. Construct houses with strong foundation and sufficiently high above the ground.

4. Leave enough space in between the buildings so that flood water can flow around the houses. If there is no free space the water will search its way, may be through the buildings.
5. Ensure that there are safe places at a short distance for rescue operations.

Floods are usually managed through a series of activities primarily to tame the overflowing rivers, not to let the excess run-off water resulting from heavy rains into the drainage network, keep flood water away from the people, provide artificial cut-off to individual bands or a series of bands so that there may be rapid flow of water in downstream. Flood water may be reduced by construction of flood control storage reservoirs. Diversion channels should be constructed so that flood water may be drained to low lying areas, depressions, etc. The disastrous effect of flood may also be reduced by improving embankment, by building of artificial levees of earthen materials, stones or even concrete walls. The key issues in management of floods which are of paramount importance are forecasting, warning and monitoring. However, it may be noted that floods are natural disasters but it depends on us how to minimize their ill effects by our technical skill, forecasting, warning, monitoring and relief and rescue operations .

Various scientists, engineers, research workers and social activists have suggested flood control measures (Chaturvedi, 1981; Das 1989; Saxena, 2005; Shukla, 2004, 2007; CWC, 1989; Kulshrestha, 2000; Sharma, 1995; Ward, 1978; Thakur, 2008; WMO, 2004, Flood Memorandum Rajasthan - 2007). World Meteorological Organization (2004) has suggested following measures for flood management in an integrated manner. These are discussed below:

### (i) Traditional Methods

Primarily the flood management practices are based on minimizing the occurrence of floods and mitigating damages caused. There are number of structural, non-structural, physical and institutional measures which may be implemented before, during and after the flood. The flood management interventions suggested by WMO (2004) are as follows :

(i) Source control to reduce run-off, for example, permanent basements, afforestation, etc.,

(ii) Storing run-off, for example, detention basins, wetlands, and reservoirs,

(iii) Increasing the capacity of river by constructing bypass channels, by channel deepening and widening of channels,

(iv) Separating the river and population, viz., proper land use planning, dikes, flood proofing and house rising,

(v) Emergency management during the flood, for example, flood warning, emergency works to raise or strengthen dikes, flood proofing and evacuation, and

(vi) Flood recovery, viz., counseling, compensation or insurance, etc.

**(ii) Afforestation**

If the soil is porous and permeable, the infiltration of flood water is considerably augmented. Similarly it happens more in those soils also which are rich in humus and are formed under deciduous forests. Shukla (2007) has observed that a dense vegetative cover not only promotes widespread infiltration but also facilitates a large amount of water to dispose off by evapo-transpiration. Tree canopy, tree roots and litter on forest floors help to break the impact of rain, bind the particles together and stabilize the forest soil and slope structures, which have been almost destroyed by overgrazing. A thick vegetative cover effectively restricts the erosive efficiency of even fiercely moving water and, thus, prevents production of countless narrow channels in the steep slope. He has further stated that termites make mounds on the abandoned stumps of trees. The termites gradually eat up the stumps and the root system and create big holes and burrows tunnels in the ground that enhances the capacity to absorb the excess run-off. The effective measure of flood management is, therefore, to bring the catchments area under dense afforestation with trees that generate heavy amount of litter.

**(iii) Detention Basins and Reservoirs**

Surface water storage by detention basins, reservoirs and by constructing small check dams are usually adopted to attenuate flood hazards. It has also been has observed that such storage serves multiple purposes and the exclusive flood storage can be first casualty in any conflict situation. Moreover, by completely eliminating the low floods such measures give a false sense of security. Storage has to be used in an appropriate combination with other structural and non-structural measures (WMO, 2007).

**(iv) Channel Improvement**

Deepening, widening and clearing away the debris and vegetation, and paving the floor increases the flood conveyance capacity of the river.

**(v) Land Use Measures**

It has been suggested by WMO (2004) that land use planning/ management is generally adopted where intensified development on a particular flood plain is undesirable. Providing incentives for development to be undertaken elsewhere can probably work better than simply trying to stop development on the flood plain. However, where land is under development pressure, especially from informal development, such planning constraints are unlikely to be effective. Flood proofing or house building are most likely to be appropriate where development intensities are low and properties are scattered or where the warning times are short. In frequently flood prone areas, flood proofing of the infrastructure and communication links can reduce the debilitating impacts of floods on economy.

**(vi) Floods Warning**

Warning of floods, alongwith forecasting, is an important consideration in flood management. Loss of human life and livestock, and property can be reduced to a great extent by giving timely warning about the coming floods. As a matter of fact, people, cattle and valuable moveable property can be shifted to a safer place. It may be pointed out that our country is one of the earliest countries to realize the importance of developing the flood forecasting and warning system in flood management measures. As such, a network of flood forecasting stations, covering almost all the States and Union Territories (UTs) was developed and, subsequently, flood forecasting and warning are being transmitted for alerting the people and all concerned agencies which are responsible for management of flood havoc (Flood Memorandum, 2007; WMO , 2004; Shukla, 2007).

**(vii) Evacuation**

Forecasting and early warning of flood hazard and evacuations are simultaneous processes. The evacuation may be upwards or outwards depending upon the conditions. The planning of evacuation may be done prior to flood hazard. It is suggested that apart from administration, non government organizations should participate actively for effective evacuation (WMO, 2004; Flood Memorandum, 2007).

## FLOOD MANAGEMENT: AN INTEGRATED APPROACH

There are two types of methods in dealing with the perennial problem of floods in India, namely, structural methods and non-structural methods.

In structural measures for flood mitigation and protection ,we construct flood control structures such as dams, embankments, and flood containing channels along urbanized areas. The structural measures, though very effective, are costly. An alternate way to deal with floods is to practice non-structural measures of floodplain management, develop efficient preparedness at district, state, and national levels, developing early warning systems, and the development and use of new technologies.

The Associated Programme on Flood Management (APFM) is a joint initiative of the World Meteorological Organization and the Global Water Partnership. It promotes the concept of Integrated Flood Management (IFM) as a new approach to flood management. The programme is sponsored by the governments of Japan and the Netherlands. The World Meteorological Organization is specialized Agency of the United Nations. It coordinates the meteorological and hydrological services of 185 countries and territories and, as such, is the centre of knowledge about weather, climate and water. The Global Water Partnership is an international network open to all organizations involved in water resources management. It was created in 1996 to foster Integrated Water Resources Management (IWRM).

For sustainable development, the principle of Integrated Water Resources Management (IWRM) was accepted in Dublin Conference (1992). It is a process which promotes the coordinated development and management of water, land and other resources, in order to maximize the resultant economic and social welfare in an equitable manner, without compromising the sustainability of vital ecosystems (WMO, 2004). Integrated flood management involves an integrated rather than fragmented approach. As such, land and water resources in relation to IWRM are utilized for maximum benefits from flood plains and minimize losses from floods. WMO (2004) has suggested that functioning of river basin should be improved as a whole while recognizing that gains and losses arise from the changes in interactions between the water and land environment and, consequently, there is a need to balance development requirements and flood losses. Side by side, it is necessary to reduce the losses from floods under integrated flood management in order to accrue maximum benefits from flood plains (Dublin Conference, 1992). A model of integrated flood management has been developed (Fig 3).

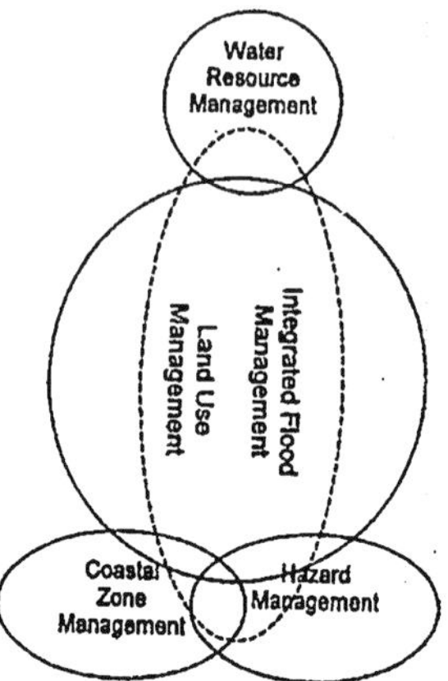


**Fig. 3 : Integrated Flood Management Model**

**Source :** WMO, 2004

Integrated Flood Management (IFM) is based on the principle of reducing vulnerability through building resilience and developing a culture of prevention through preparedness rather than reactive responses alone. Accordingly, following points should be taken into consideration for integrated flood management:

(i) Management of water cycle,
(ii) Integrate land and water management,
(iii) Adopt a best mix of strategies,
(iv) Ensure a participatory approach, and
(v) Adopt integrated hazard management approaches.

## INTER-LINKING OF RIVERS

Northern plains and North-eastern areas of our country are worst-hit by floods. Many workers have suggested a system for inter-basin transfer of flood water to water shortage areas or drought affected areas. It may be pointed out that Narmada river rises in a relatively high rainfall region in M.P., and if its abundant flood water are stored and then diverted to the areas of Gujarat most troubled by drought and salinity, such efforts may enable permanent and near-permanent solutions to floods. They further suggested that if Narmada and Godavari rivers are linked with Cauvery river, the problems of severe floods can be tackled. Considering the long-term benefits and to solve the dual problems of floods and droughts, it is necessary that all the surplus water from rivers is diverted to the water deficit or drought prone areas. As per estimates, such net working will not only control the devastating floods and recurrent droughts of the regions, but the water will also be used to produce about 450 million tonnes food grains needed by 2050 AD for the projected population of 164 million people. As such no water will go waste to the sea, ultimately avoiding devastating floods (Reports of National Commission on Floods, 1980; Shukla, 2008).

Various plans of inter-linking of Indian rivers have been put forward, but due to widespread problems of feasibility, desirability and viability these plans are still under wide discussions. In 1990, the government of India appointed a commission to examine the strategy of water resource development, including the possibility of interlinking rivers. Later on, in 2001 a task force was set up by government of India, as per the directions of Supreme Court, which has directed the centre government to implement the programme of inter linking of rivers by 2015. It is estimated that this ambitious project to link 30 rivers and transfer surplus water to drought prone areas is likely to cost Rs. 5,60,000 crores and is to be completed by 2016. The proposed inter-basin water transfer links are shown in Fig. 4. Yadav (2003) has proposed to go for inter-basin transfer of water from surplus to deficit areas as envisaged in National Perspective for Water Resources Development.

The idea of interlinking of rivers in India has repeatedly been occurring to the Indian scientists and engineers in the past but has always met with rejection on account of technical unfeasibility or socio-economic issues.

Four decades back, it was Mr. K.L. Rao, the then Union Minister for Irrigation and Power, who suggested a 2640 km long Ganga-Cauvery link by carrying waters partly by gravity and partly by lift but the costs were found highly prohibitive and the idea was rejected. This was followed by the Captain Dastur Plan, popularly known as Garland Scheme. Under this scheme, a 4200 km long Himalayan Canal and a 9300 km long Garland Canal were suggested, both to be linked by pipelines near Patna and Delhi. Again, the plan was found technically infeasible and laid to rest.

Presently, the Ministry of Water Resources, Government of India, has come up with an elaborate plan of linking various rivers to mitigate droughts and floods for maximum utilization of river water that is going waste into the sea. Supreme court has desired the study of feasibility of the scheme and there is a mixed response from different corners of the country.

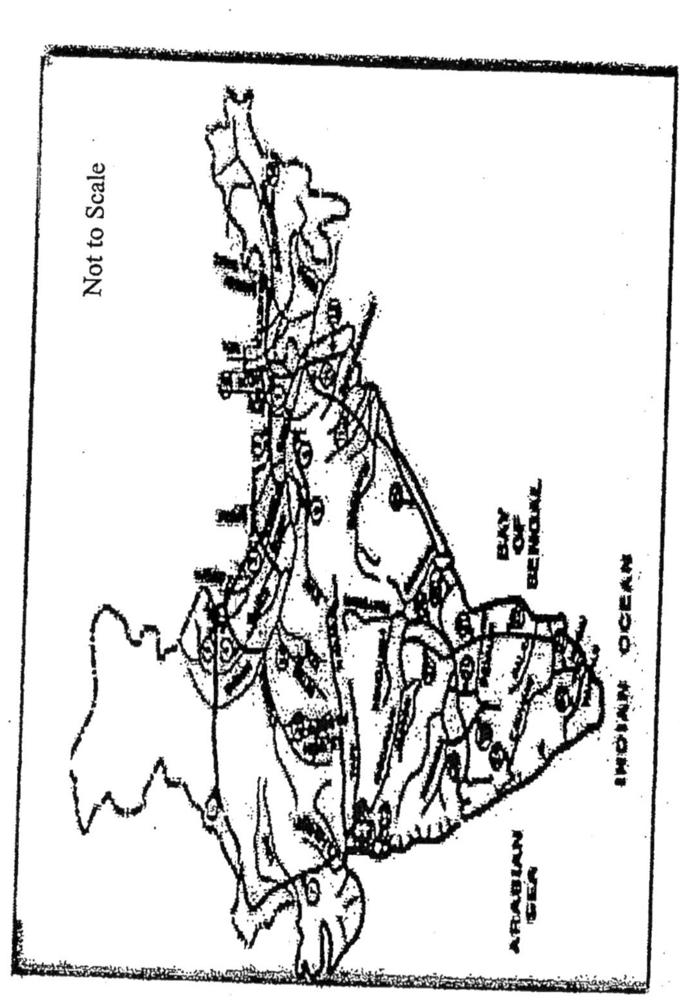


**Fig. 4 : Proposed inter basin water transfer links**

**Source:** National Water Development Agency, Ministry of Water Resources, Govt. of India, 2002

As per the newly proposed scheme, two components, namely Himalayan component and Peninsular component have been outlined. The Himalayan component has been proposed to have 14 canal links (Table 3).

**Table 3: Links identified in Himalayan and Peninsular component**

| *S.N.* | *Himalayan Component* | *Peninsular Component* |
|---|---|---|
| 1. | Kosi - Mechi | Mahanadi (Manibhadra) -Godavari (Dowlaiswarm) |
| 2. | Kosi - The Ganges | Godavari (Inchampalli)-Krishna (Nagarjunsagar) |
| 3. | Gandak - The Ganges | Godavari (Inchampalli Low dam) - Krishna (Nagarjuna Tail Pond) |
| 4. | Ghagara - Yamuna | Godawari (Polavaram) Krishna (Vijaywada) |
| 5. | Sarda - Yamuna | Krishna (Srisailam) - Penner (Prodattur) |
| 6. | Yamuna - Rajasthan | Krishna (Srisailam) - Penner (Prodattar) |
| 7. | Rajasthan - Sabarmati | Krishna (Nagarjunasagar) - Pennar (Somasila) |
| 8. | Chunar- Sone Barrage | Pennar (Somasila) - Cauvery (Grand Anicut) |
| 9. | Sone dam - Southern tributaries of the Ganges | Cauvery (Kattaiai) - Vaigai -Gundar |
| 10. | Brahmputra - The Ganges (Manas - Sankosh - Tista- The Ganges) | Ken-Betwa-Link |
| 11. | Brahmputra - The Ganges (Jogigopa - Tista- Farakka) | Parbati - Kalisindh- Chambal |
| 12. | Farakka - Sunderbans | Par-Tapi-Narmada |
| 13. | Farakka - Damodar - Subarnrekha | Damanganga - Pinjal |
| 14. | Subarnrekha-Mahanadi | Bedti - Varda<br>Netravati - Hemavati<br>Pamba - Achankovil - Vaippar |

Under the Himalayan component, more dams are planned to be constructed on the tributaries of the Ganges and Brahmaputra in India, Nepal and Bhutan. This component envisages linking of Brahmaputra and its tributaries with the Ganges and the Ganges with Mahanadi to transfer surplus flow from east to the west, thus benefiting many states. The proposed links may help in controlling floods in the Ganges and Brahmaputra basins.

Under the Peninsular component, surplus waters of Mahanadi and Godavari rivers are proposed to be transferred to the deficit basins of Krishna, Pennar, Cauvery and Vaigai. In the transfer of waters from Godavari to Krishna, a lift of 1200 cusecs of water over about 116 m is essential. In addition, water is proposed to be transferred from Ken river to Betwa river to benefit M.P. and A.P., interlinking of Parbati, Kalisindh and Chambal rivers to benefit M.P. and Rajasthan. Peninsular component aims at benefiting Orissa, Karnataka, Tamil Nadu, Pondicherry, Maharashtra and Gujarat.

Estimated cost of the whole scheme at 2002 price level has been worked out to be Rs, 5,60,000 crore, and the proposed time period for the completion of the scheme is 35 years. Whether the linking of rivers will promote integration or generate more disputes and tension is a moot question (Vaidyanathan, 2003; Singh, 2003). The present status of implementation of the scheme is that the feasibility report of six links has been completed while that of 18 links is under progress (Goyal, 2003).

Besides several obvious, but *prima facie,* important questions about the concept, feasibility, desirability and viability of the proposal need to be clarified before its implementation. It appears that interlinking of rivers may not become a big dream of little logic. However, Mr. Suresh Prabhu, Chairman, Task Force on Interlinking of Rivers, has observed that in view of the large variations in rainfall over space and time, our country experiences frequent floods in some parts and severe droughts in other parts. Floods are frequent features particularly in the Brahmaputra and the Ganges rivers, which carry 60 per cent of the water resources of our country. Flood damages, which were of the order of Rs. 52 crores in 1953, went upto 5846 crores in 1998 with an annual average of Rs 1343 crores, affecting Assam, Bihar, West Bengal and Uttar Pradesh, causing untold human miseries. On the other hand, large areas in Rajasthan, Gujarat, Andhra Pradesh, Karnataka and Tamil Nadu face recurring droughts. It is expected that interlinking will provide additional irrigation benefits to 35 million hectares of agricultural land. He has further suggested that the link canals will be 50 to 100 m wide and more than 6 m deep. This would facilitate inland navigation from North to down South. A boost to fresh water fisheries is also expected as a result of the programme. Apart from these benefits, guaranteed minimum flows in the rivers will strengthen ecology and health of environment. Hydro-electric power could also be generated on a massive scale by the storage dams. As per the estimates, against the potential of 84000 MW only about 22000 MW capacities for hydro-power generation has been developed so far. For an efficient working of electrical energy generating system, the mix of thermal to hydro should be 60:40. The storage dams proposed under interlinking of rivers will greatly improve the situation. The total hydro-power potential of the interlinking systems is assessed to be 34,000 MW (Singh, 2003). Apart from other benefits, drinking water availability will also improve substantially.

## CONCLUSION

The distribution of flood prone areas in our country has been described in detail. As a matter of fact, India shares 16 per cent population where as its fresh water resources are only 4 percent. As such, it is utmost important to conserve rain water which goes waste as floods every year. It will not only save the people from disasters but the flood water will be utilized for various purposes. While it is impossible to control nature, human activity, its impact on the environment and subsequent impact on vulnerability can be altered. A mechanism for sharing the costs and benefits of flash flooding mitigation works should be devised. Thus, the integrated flood control measures are of great importance. Large scale plantation, proper planning and implementation of inter- basin transfer of flood water without disturbing ecosystems are some of the key measures for mitigation of floods. People's participation in various rescue operations is also necessary. People should also be educated about such dangers. For successful implementation of flood control measures, a group of concerned scientists, meteorologists, disaster management experts, administrators, policy makers should be created in each state headed by a very senior officer so as to coordinate among various departments to manage the floods and to make best utilization of flood water, that goes waste, for human and livestock consumption as well as agricultural and recreational purposes.

There is an immediate need to monitor the weather parameters, and run-offs from streams and river, and installation of early warning system in disaster prone areas.

## REFERENCES


Anonymous (1980). Report of the National Commission on Floods. Ministry of Energy & Irrigation, Govt. of India, New Delhi, pp. 2-20.

Anonymous (1989). *Manual of Flood Forecasting.* Central Water Commission, New Delhi, pp. 1-15.

Anonymous (1992). *Integrated Water Resource Management.* Dublin Conference, WMO, Geneva, Switzerland, p. 1.

Anonymous (2003). The Independence Day Feature. *The Hindu,* An English Daily, 11th Aug., pp. 1-3.

Anonymous (2004). *Integrated flood management : A Concept Paper.* Technical document No. 1, APFM, World Meteorological Organization, Geneva, Switzerland.

Anonymous (2007). Flood Memorandum. Water Resources, Govt. of Rajasthan, Jaipur, pp. 104-114.

Anonymous (2008). Rs. 1000 Crore for flood hit Bihar - P.M. *Times of India,* 29th Aug., p. 1.

Anonymous (2008). *Times of India.* 29th August, 2008.

Chandra, Suresh (2003). *Flood Management in Damodar River Basin: A Case Study.* Integrated Flood Management World Meteorological Organization, Geneva, pp. 1-7.

Chaturvedi, M.C (1981). *Flood Management : New Concepts, Technology & Planning Approach.* Indian National Science Academy, New Delhi -15.

De, V.S, Dube, R.K, Prakash Rao, G.S. (2005). Extreme weather events over India in last 10 years. *J. Indian Geophys. Union,* 9(3) : 173-187.

Ghani, M.U. (2003) Participatory strategy for flood mitigation in East & North-east India: A case study of Ganges - Brahmaputra - Meghna basin. Unpublished Report (http://www.unescap.org/esd/water disaster/200 1/india.doc).

Goyal, Jagvir (2003). Interlinking of Rivers : Facts & Figures. *The Tribune,* March 13,.2003.

Naithani, A.K. (2003). Flash floods in the Himalaya. *Employment News,* 28(5): 1-2.

Niwas, Ram and Khichar, M.L. (2010). Weather hazards and their mitigation. Chapter 9, *In : Natural and Man Made Disasters: Vulnerabilty, Preparedness and Mitigation* (Eds. K.K. Singh and A.K. Singh). M.D. Publications Pvt. Ltd., New Delhi, pp. 229-252.

Saxena, N. (2005) Flood : The Indian Scenario. *Employment News,* 30 (21): 24.

Sharma, J.R, Bothale, R.V, Dutta, D, Paliwal, R, Bothale, V.M, Gupta, A.K., Bera, A.K, Bhadra, B.K, Banu, V, Gaikwad, R.N, Sreenivasan, K.(2006). Floods in Barmer District, Rajasthan : An analysis based on Remote Sensing and GIS. *34th R.G.A. National Conference on Environmental Conservation & Resource Management.* P.G. Dept. of Geography, Govt. Dungar College, Bikaner.

Sharma, V.K. (1995). *Disaster Management.* Indian Institute of Public Administration (IIP A), New Delhi, pp. 2-10.

Shukla, M. K. (2007). Flood and the lasting remedy. Chapter 5, *In: Water Resource Management and Development* (Eds. K.K. Singh *et al).* Kalyani Publishers, New Delhi, pp. 86-106.

Singh, B. (2003). Interlinking of rivers: A big dream of little logic. Chapter 6, *In : Water : Problem and Its Management.* Hope India Publications, Gurgaon, pp. 61-65.

Sinha, A. (2008). Did Nepal neglect lead to Kosi floods. *Times of India,* Aug. 29, p. 6.

Thakur, D.K. (2008). Environmental hazards and disaster. Chapter 1, *In: Natural Hazards and Disaster Management* (Ed. B.C. Jat). M.D. Publications Pvt. Ltd., New Delhi, pp. 1-62.

Vaidyanathan, A. (2003). Interlinking of rivers : Can it solve India's water problem. Chapter 6, *In : Water Problem and Its Management.* Hope India Publication, Gurgaon, pp. 47-60 .

Verma, N.K. (2008). Developing on Asian Tsunami Warning System. Chapter 6, *In: Natural Hazards and Disaster Management* (Ed. B.C. Jat). M.D. Publications Pvt. Ltd., New Delhi, pp. 101- 107.

Ward, R. (1978). *Floods - A Geographical Perspective.* The McMillan Press Ltd., New York, p. 7.

Yadav, Satish (Ed.) (2003). *Water - Problem and Its Management.* Chapter 2, *In: Water Problem and Its Management.* Hope India Publication, Gurgaon, pp. 23-29.

## Chapter 6

# Impact of Droughts in Rajasthan

*S.K. Mathur*[1] *A.K. Srivastav*[2], *Surjit Kaur*[3] *and H.C. Sharma*[4]

### INTRODUCTION

Drought is a major natural disaster, particularly in desert areas of Rajasthan. Inadequate, unreliable and uneven distribution of rainfall creates the situation of drought in the area. One can neither prevent failure of rain nor can even put forward any claim to produce timely and adequate rainfall to the sweet desire of agriculturists. As such, crop production gets drastically reduced, migrations is forced, lot of livestock perishes and entire economy of masses, particularly in rural areas, frequently get shattered (Mathur, 1981 : Chouhan, 1984). Severe drought also result in depletion of surface and ground water levels; soil erosion leading to loss of fertile soils, dust storm, dust hazes, triggering sand dune movement, blocking of roads, rail tracks, crop fields, thus, creating excessive stress on human and livestock population. The studies conducted in arid region indicated that meteorological droughts occurred in one area or the other, in 30 55 per cent of the years, during 1901 to 2000, indicating thereby that every third year is a drought year. They may also persist continuously for more than one year, as has been experienced during 1962 to 1970 and 1984-1987 (Raj., 2004). As per UNO report (2007), during 1991-2000, more than 2,80,000 persons died due to drought in the whole world.

As far as definition of drought is concerned, it may be defined in many ways. Oliver and Fairbridge (1987) suggested that drought is :

[1]Quality Control Laboratory, Krishi Bhawan, Bikaner (Raj.)-334001, India.

[2]Project Directorate (Research), Agriculture & Soil Survey, Krishi Bhawan, Bikaner (Raj.) - 334001, India.

[3]Rashtra Sahayak Sr. Secondary School, Bikaner (Raj.)-334003, India.

[4]Office of Agricultural Extension, Public Park, Bikaner (Raj.)-334001, India.

(i) a period of rain deficiency,
(ii) a relative state of forest flammation,
(iii) occurring when a specific agriculture crop, or pasture, yields less than expected amounts, and
(iv) denoting a critical level of soil moisture or ground water depletion, etc.

## FACTORS OF DROUGHT

Physiography, climate, soil condition, rainfall, population inflammation, lack of infrastructure, ignorance of traditional and small scale industries, high dependency on agriculture, short term relief operation and ecological imbalance are the main reasons which lead to drought conditions (Jhalani, 2004) as depicted in Fig. 1.

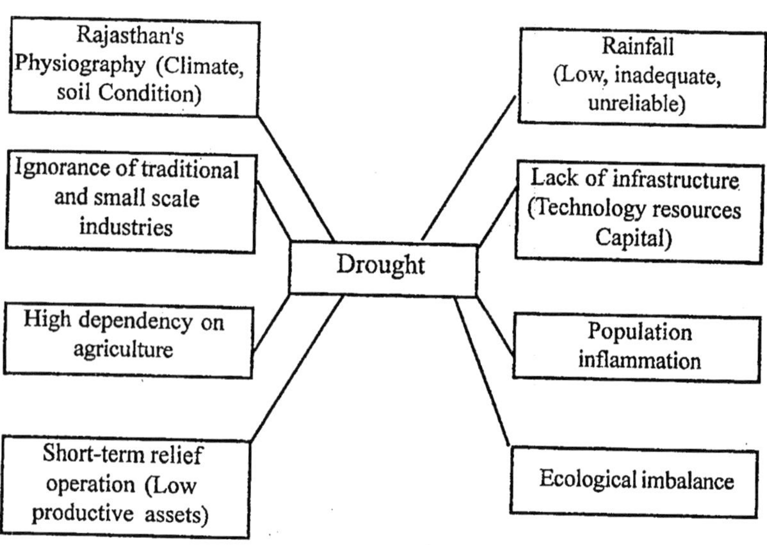


**Fig 1 : Causative factors of drought**

Meteorologists have given several partial explanations of droughts in our country, based on theoretical grounds, as well as on the past synoptic situations related to monsoon failures. Main factors are as under:

1. Weaker meridional pressure gradients,
2. Larger northward shifts of the monsoon trough,

3. Larger number of breaks in the monsoon, and
4. Smaller frequencies of depressions, and shorter westward extent of depressions tracks.

The studies have suggested that there are remote meteorological phenomena which are correlated with, if not causative of, the droughts in India. The *El Nino* phase of the Southern Oscillation (ENSO) has impacts over drought in India. The location of quasi-stationary anticyclonic circulation over north-west India has been found to inhibit rain producing conditions. Shankar Rao and Sachidananda (1987) have come out with three major hypotheses about drought mechanics:

1. SST- based Walker communication hypothesis (SSTWTC)
2. Tropical internally forced or free oscillation hypothesis (TIFFO), and
3. Latitudinal teleconnection hypothesis (LTC)

It may be pointed out that if mid-latitude longwave systems remain quasi-stationary, then they result in stronger teleconnections through the tropospheric monsoonic window region around 80°E. When the phase is right, monsoon breaks.

## CLASSIFICATION OF DROUGHTS

As per Meteorological department, drought is a situation occurring in any area when the annual average rainfall is less than 75 per cent of the normal. A rainfall of more than 25 per cent, but less than 50 per cent of the normal is termed a moderate drought, whereas a severe drought is one where the deficit exceeds 50 per cent. A drought-prone area is one where at least 20 per cent of the years have recorded deficit rainfall. If this percentage exceeds 40, the area is designated as a chronically drought-prone area (Indian Irrigation Commission, 1972). Sarkar (1979) computed the total areas which received less than 75 per cent of the normal monsoon rain for a run of years, and expressed these areas as percentage of the total area of India. The years in which the area exceeded 25 per cent were designated as drought years. Bhalme and Mooley (1980) considered the variability of rainfall and developed a numerical drought index (Im or monsoon index, for -1.00 to -4.00 or less) based on monthly monsoon rainfall and its intensity of drought. The drought area index (DAI) is defined as the percentage of the area of India having a mean index of –2 or less. If the DAI is 25 or more in a year, it is defined as a drought year. The National Commission on Agriculture (1976) classified drought as meteorological, hydrological, agricultural, ecological, socio-economic drought and famine.

### (a) Meteorological Drought

As stated above, the Indian Meteorological Department has defined meteorological drought as a situation when annual rainfall over an area in a year is less than 75 per cent of the normal. An area is termed as drought area if it experiences drought conditions in 20 per cent of time, whereas if the frequency is more than 40 per cent in certain area, it is termed as chronically drought prone. Most of the Rajasthan, along with Kutch, falls under this category (NCA, 1976).

### (b) Hydrological Drought

Prolonged meteorological drought results in hydrological drought with a marked depletion in surface water and consequent drying up of reservoirs, lakes, streams and rivers, cessation of flows and fall in the ground water level. Subrahmanyam and Subrahmanyam (1964) found that aridity index, viz., per cent ratio of annual water deficit to annual water need, proposed by Thornthwaite (1948) was the most useful parameter to classify droughts in terms of standard deviations for normal aridity index. Later on, Sastri and Ramakrishna (1980) modified the explanations to classify the droughts in arid regions (Table 1).

**Table 1 : Classification of droughts in arid region**

| *Intensity of drought* | *Category of drought* |
|---|---|
| 0 or above | No drought |
| 0 to -0.5s | Mild drought |
| -0.5 to -1s | Moderate drought |
| -1s or below | Severe drought |

s - Signifies standard deviation

**Source:** Ramakrishna, 1980

### (c) Agriculture Drought

Agriculture drought means the difference between the water requirement (Evapotranspiration) and water availability during crop season. As such, the intensity of drought depends upon water stress which occurs during different stages of crop growth. The intensity of agriculture drought can be determined if water stress during growth period is evaluated on the basis of water requirement. Rao *et al.* (1981) have classified agriculture droughts on the basis of AE/PE ratios (actual evapotranspiration/potential evapotranspiration) during different stages of crop growth.

**(d) Ecological Drought**

The ecological drought occurs when the productivity of natural ecosystem fails substantially due to environmental damage caused by distress.

**(e) Socio-economic Drought**

The socio-economic drought occurs when production of food grains fails and a situation of acute food shortage prevails in the affected area.

**(f) Famine**

Famine situation occurs when there is a large scale collapse of access to food, which results in mass starvation unless there is some external assistance. As such, there is a great distress and sufferings with large scale migration and dying of livestock.

**Drought: Indian Scenario**

Various studies have been conducted by number of scientists regarding drought conditions in our country. The worst droughts in the pre-independence period were recorded in the year 1877 and 1899 when rainfall departures were - 79% over 66.8% of the area, and - 26.2 over 83% of the area, respectively. Both these droughts caused severe famine conditions. After independence, there has been no serious famine but droughts have occurred now and then. In the recent years, country has faced three severe droughts in 1979, 1982, and 1987 (Sarkar, 1979; Das, 1984, 1988).

About 19 per cent area, and 12 per cent population, is affected by drought in India. These areas receive low to very low rainfall. It has been mentioned in the irrigation commission report (1972) that all those areas, receiving less than 75 cm of rainfall, are drought prone. The indiscriminate use of ground water has caused the water table to go deep down. About 70 – 80 per cent farm produce is attributed to well-water irrigation. This situation is also responsible for recurring droughts. The drought prone areas of our country are demarcated in Fig. 2.

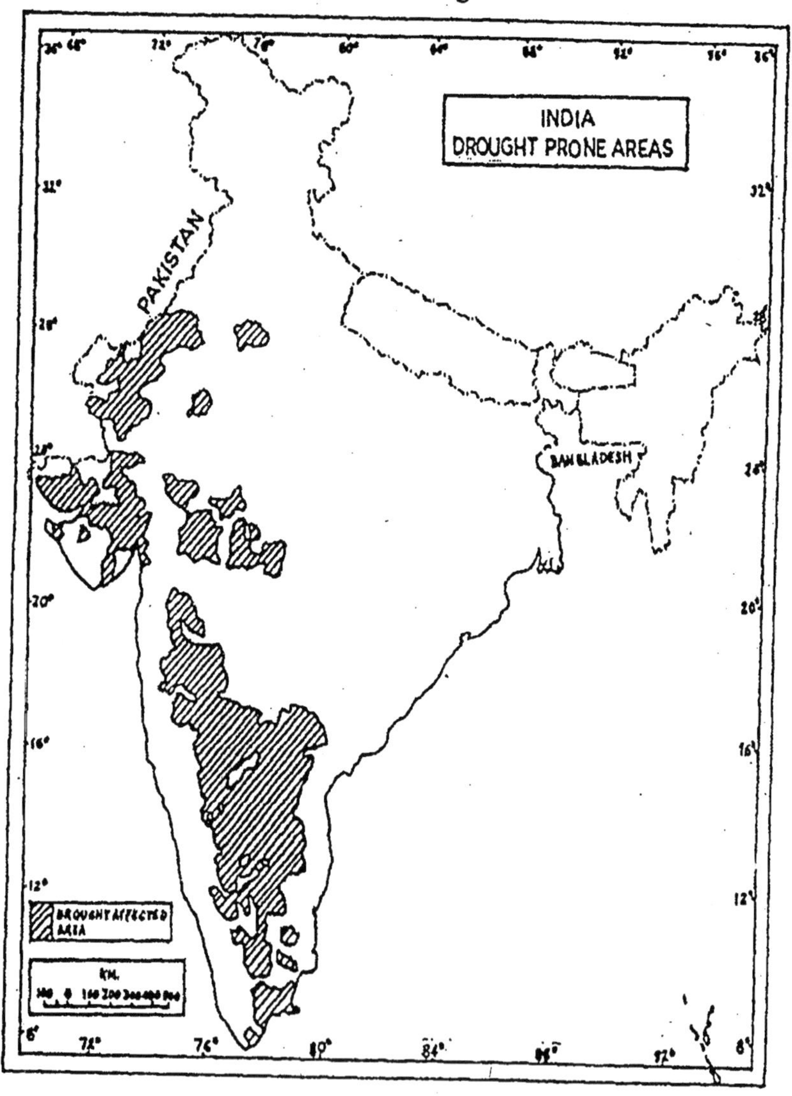


**Fig. 2. : Drought Prone Areas of India**

## IMPACT OF DROUGHTS IN RAJASTHAN

By and large, drought is associated with scarcity of water. However, for agricultural people, it means deficit of rainfall, and of soil moisture, which prevents good crop growth. For a meteorological, drought indicates very low to low rainfall, as compared to normal one in a particular area, and to a hydrologist it indicates the deficit of water in surface and ground water resources. Geographically, Rajasthan comes under arid (rainfall upto 375 mm) and semi-arid zone (rainfall 375-750 mm) of the country, and >90% of the rainfall in this region is received during south-west monsoon. The monsoon period is not one of continuous rain in any part of the state. The impacts on various aspects are discussed as follows:

### 1. Crop Production

It has been observed that, in western Rajasthan, 35-48 per cent yields of pearlmillet were reduced from the normal yield due to moderate drought conditions that prevailed in the year 1979-82 and 1984-85, whereas in the severe drought conditions in the year 1987, the reduction in yield was recorded as high as 93 per cent. Ramana Rao *et al.* (1984 a & b) and Ramakrishna *et al.* (1988) have observed that moisture availability conditions during 6$^{th}$ to 9$^{th}$ week after sowing have a strong influence on the grain yield of green gram and cowpea. For pearlmillet, water availability conditions during 7$^{th}$ to 11$^{th}$ weeks were found to influence ultimate yields, while in sesame the water availability during 6$^{th}$ to 10$^{th}$ week correlated with the grain yield. The probabilities for occurrence of various intensities of drought under pearlmillet, studies by Sastri *et al.* (1982) are presented in table 2. The results indicated that Sikar is free from droughts for a maximum period (for 75 per cent of the years), followed by Barmer (for 53 per cent of years). It was also observed that probability for drought occurrence increases from 23 per cent at Sikar to 88 per cent at Jaisalmer. The results of the study further revealed that pearlmillet experienced no drought conditions in 31, 16, 7, 7 and 1 per cent years in Sikar, Jodhpur, Nagpur, Barmer and Jaisalmer districts, respectively. As regard the yield levels, the data indicated very low yields in Sikar (2.9 $qha^{-1}$) and Jaisalmer (1.1$qha^{-1}$).

**Table 2 : Probabilities for occurrence of agricultural droughts under pearlmillet in western Rajasthan**

| *Station* | *Mean Annual Rainfall (mm)* | *Percentage Probability of Agriculture Drought* | | | |
|---|---|---|---|---|---|
| | | *Nil* | *Mild* | *Moderate* | *Severe* |
| Sikar | 468 | 31 | 44 | 3 | 22 |
| Jodhpur | 368 | 16 | 28 | 16 | 40 |
| Nagaur | 332 | 7 | 43 | 11 | 39 |
| Barmer | 260 | 7 | 46 | 7 | 40 |
| Jaisalmer | 189 | 1 | 11 | 9 | 79 |

**Source :** Sastri *et al.*, 1982

**2. Livestock**

The worst victims of drought happen to be the animals and cattle, who are major components of life supporting system to the population in the otherwise drought-prone area. The impact of drought on livestock may be discussed in four ways : (i) mortality, (ii) loss in productivity, (iii) health of animal, and (iv) loss of fertility. Rathore (2005) has observed that in drought year 2003, the populations of sheep and camels, followed by cows were most vulnerable amongst animals. They perished in large number, ranging between 10-32 per cent of the population. The data further indicated that loss of cows was much higher as compared to state census data, as large number, i.e., 25 to 50 per cent of cows was abandoned to starve because of acute shortage of fodder and finance. The remaining cows became weak. There was nobody to buy cows even for slaughter, as there is a ban on cow slaughter in Rajasthan. The severity of drought can be judged by the fact that even high price animals like buffaloes suffered and few perished. It was further assessed by Rathore (2005) that goats and sheep are also affected by drought, but relatively mildly, as compared to large animals. They are mainly grazing animals and depend on common property resources, i.e., grazing lands, wasteland, forest etc. Moving to better-endowed areas saves them. Migration of small animals is traditional practice in Rajasthan. The drought year of 2003 was bad for sheep as there was widespead drought in Rajasthan and also in adjoining states, therefore, even the migration strategy did not work and more than 32 per cent of sheep perished.

**3. Human Population**

In fact human population is also badly affected by drought. Food availability is reduced as crop production fails. Side by side, income of the people employed in agriculture sector is reduced, and these people are forced to change their occupation. As such, people migrate from their villagers to other places in search of jobs, and in such situations people are forced to either sell their assets or borrow money. Rathore (2005) has reported that change in occupation and migration in 2003 due to drought in Rajasthan had a significant impact on the occupations. About 80 per cent farmers were forced to join the labour force, both at relief sites and outside villagers.

**4. Ecology**

Drought has ecological effects through inappropriate and overextended use of environmental resources like land, water and forest. Wind erosion from bare and dried up soil, and reduced infiltration of rain water into the soil in the successive rainy season, coupled with increased surface run-off, are the adverse for desertification, and desertification cycle is most important as far as ecological changes are concerned. Bose and Saxena (2000) have concluded that impact of drought cannot be assessed only on the basis of economic loss. However, social and ecological damages are irreversible, and as such, their impact persists for years together.

## DROUGHT MANAGEMENT IN RAJASTHAN

Various efforts made by way of short term famine relief operations, and thousand of million rupees spent to combat the famine hazard during and prior to 4th Five Year Plan, could not solve the problems and give relief on permanent basis. This necessitated adoption of an integrated rural development approach to conserve, develop and utilize the available land, water and livestock resources of desert area on scientific basis, so as to mitigate the effects of drought and develop stable economy. Jhalani (2004) has suggested drought management plan in Fig. 3.

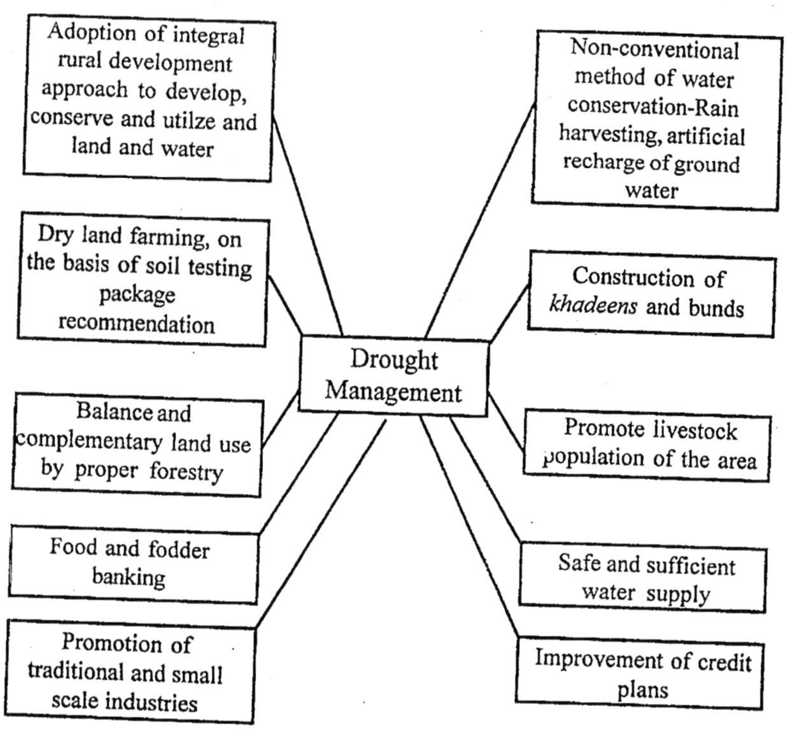


**Fig. 3: Flow chart of drought management**

Various development programmes of region wise suitability, based on necessary survey of resources were, therefore, formulated and their implementation started from the beginning of 5th Five Year Plan under Drought Prone Area Programme (DPAP), and later on under Desert Development Programme (DDP), Integrated Rural Development (IRD) and *Antyodaya* Programme with the following objectives :

1. To reduce the extremity of drought and control of desertification,
2. Increase and stabilize the economy of masses, particularly weaker sections,
3. Restoration of ecological balance, and
4. Integrated rural development for the creation of rural employment, through construction of productive assets and individual beneficiary scheme.

The above mentioned programme are being implemented in the State, with financial assistance obtained from state government as well as government of India. These programmers are administered by Commissioner and Secretary of Agriculture Department, Special Schemes Organization, Jaipur and assisted by Deputy Secretaries and some technical experts for the purpose of planning, budgeting and supervision. A state level coordination committee, comprising of concerned Heads of Departments and Secretaries, periodically reviews the progress under the Chairmanship of Chief Secretary. At district level, District Development Agency (DDA) has been formed for coordination, planning, monitoring progress and release of funds to various development departments which are involved in the execution of programme. However, at the district level, apart from administrative officers, technical officers of various departments, local representatives of *Panchayati Raj* institutions take part in the implementation programmes. Various development programmes under progress are discussed briefly as follows:

**1. Agriculture**

Based on soil, agro-economic and topographical surveys, shelf of project are prepared and executed for soil and moisture conservation so that land and water resources could be utilized on scientific lines, and higher level of production is achieved. Watershed development projects and dry land farming demonstrations are conducted so that latest technology is transmitted to the farmers. Soil testing facilities are being made available for balanced use of fertilizer in order to maintain good soil health.

**2. Animal Husbandry**

The economy of rural masses in Rajasthan is primarily based on both agriculture and animal husbandry. The later predominates in vast arid and semi-arid tracks. Out of total geographical area of 329 million ha, about 11 million ha land is estimated to be falling in various categories of wastelands. These lands, by and large, are partly or completely devoid of natural vegetation due to over exploitation and therefore, unable to provide adequate fodder to meet the requirements of huge livestock population of state which is estimated to be more than 23.4 million. Temporary migration of livestock from the area is regularly resorted to. Large scale import of fodder, mostly wheat, straw, having very poor nutritional value, has become a permanent feature under drought conditions. Similarly, the availability of fuel wood is not much different from that of fodder. With the increasing demand of both fodder and fuel due to ever increasing population of livestock and human being, the gap between their availability and demand is widening day by day and adversely affecting the rural economy (Mathur and Mathur, 1989). In order to combat the drought, we have to produce adequate quantity of fodder as well as check desertification. As such various short-term and long-term strategies are summarized as under:

***(a) Short Term Steps During Drought Period***

(i) Large scale supply of fodder crops minikits to farmers free of cost of 0.25 ha plot demonstrations,

(ii) Supply of improved varieties of fodder seeds at subsidized rates,

(iii) Extra water allowance from canals to encourage fodder cultivation,

(iv) Supply of cattle feed and fodder on subsidized rates,

(v) Feeding non-conventional and indigenous feed and fodder,

(vi) Arranging urea-molasses bricks for animals to lick,

(vii) Feeding straw treated with urea, molasses, etc.,

(viii) Organize cattle camps for disowned animals, and

(ix) Arrangement of medical aid, drinking water and fodder, particularly in camps and route of migration.

***(b) Long Term Planning***

(i) Cultivation of fodder crops by each farmer is the most important step. The farmer should be trained in this direction and may be motivated for fodder production,

(ii) Silvi-pasture on community and land should be developed under weather suitable conditions. There is need for greater participation of rural masses, and community lands may be allotted to individual farmers, cooperative societies and voluntary organizations for the purpose on attractive terms, including ownership of the produce,

(iii) Similarly, silvi-pasture may also be developed on farmers' field. Farmers should be educated and encouraged to put as much area under silvi-pasture as possible. Agro-forestry may be adopted in a comparatively assured rainfall areas, and

(iv) Fodder banks may be established, and it should be assured to purchase it at remunerative price at the time of sale, and

(v) Breed improvement programme, and elimination of unproductive animals, also needs immediate attention with ultimate objective of raising livestock on commercial lines. Animals which are uneconomical debilitated and not true to the breed need to be discarded, and government may take suitable action for their disposal and/or maintenance.

**3. Ground water**

Ground water in western Rajasthan generally occurs under water table conditions (unconfined) in hard crystalline and alluvial formations. The depth of water table ranges from 100 to 140 m below the land surface. The quality of water in the area is generally saline to highly saline. Hydro geological survey by the Rajasthan Ground Water Department is in progress. There are indications of substantial exploitable surplus ground water potential in some regions. The Department is also involved in drilling tube-wells for drinking as well as irrigation purposes.

**4. Irrigation**

On recommendation of National Commission on Agriculture, and having seen the utility of moisture conservation through *khadeens* in various desert districts, greater emphasis is now being given to the construction of *khadeens*. Irrigation works, viz., construction of bunds, etc. for water conservation are being carried out in a big way.

**5. Forestry**

Under this sector various programmes, viz., sand dune stabilization, shelter belt plantation, road side plantation, canal side plantation, silvi-patoral plantation, farm forestry, mixed plantation, block plantation, village fuel wood, fodder plantation, pasture development, etc., are being executed by forest department.

It is proposed to establish and develop a National Desert Park, spread over in an area of 4000 km$^2$ in Barmer and Jaisalmer districts with the object to conserve and develop desert fauna and animals. It is proposed to construct roads and tanks to store rain water, desert lodge, watch towers at appropriate places, and provide a tourist sand baggy, specially designed to attract tourists, and to take them to inner parts of the park.

**6. Power**

In addition to normal remunerative programmes of rural electrification of Rajasthan State Electricity Board (RSEB), power supply is also being executed by the board under DPAP and DDP. This will help in the development of those villages also which are presently considered to be non-remunerative.

**7. Cooperative and District Credit Plan**

To meet the increasing demand of short, medium and long term credit needs of masses for agriculture, animal husbandry and rural artisan works, central cooperative banks are being strengthened. District credit plans are being prepared by the concerned Lead Banks in each district.

**8. Integrated Rural Development**

This plan was initially implemented in 112 *panchayat samities*, selected in Rajasthan with the object of creating employment opportunities and training rural masses, particularly weaker sections. This is, basically, an individual beneficiary programme, and covers and all those whose source of incomes is from agriculture, animal husbandry and rural artisan works.

**9. Antyodaya Yojana**

Under *Antyodaya* Programme, which was started in Rajasthan for the first time in country in 1977, poorest of the poor families in the villages are selected on the basis of certain fixed criteria for providing them means of livelihood. Among the important items by which these poor families are benefited are: (i) old age/disabled pension @ Rs. 40/month, (ii) allotment of suitable land for cultivation, (iii) employment, (iv) distribution of *charkha* etc., and (v) advancement of loan at low rate of interest, and release of subsidy for purchase of cow, camel/camel cart, bullock/bullock cart, sheep and goats, etc.

## SUGGESTIONS FOR IMPROVEMENT

(i) Fodder banking programmes may be given maximum importance, particularly in arid areas, so as to meet fodder needs at the time of famine.

(ii) Intensive pasture development programme should be taken up. Side by side, tree plantation may also taken up on large scale.

(iii) There is need for strengthening and reorganization of block level machinery for more efficient and effective implementation of various development programmes.

(iv) More involvement of commercial banks is desired for meeting out loan needs of rural area.

(v) The uncommand areas of Indira Gandhi Canal, particularly left bank areas up to 50 km may be covered on large scale for pasture development, and creation of water points for supply of drinking water. For this purpose, water may be lifted from main canal at appropriate places.

(vi) Land policies have encouraged land use and management pattern, which is not in keeping with requirements of arid land. Landuse, according to its capabilities, with adoption of conservation technique is essential in arid region. Norms of leaving pasture land under Rajasthan Land Utilisation Act (*Gocher Bhumi* @ 0.10ha/cattle) in the state do not take entire livestock population into account, as well as carrying capacity of pasture which varies from region to region. Norms, therefore, need to be revised, and more area be made available for pasture in arid regions. Village pastures management is left with *panchayats* which does not take any step for its improvement.

## CONCLUSION

The periodic droughts in our country, and particularly in Rajasthan, are due to peculiar geo-meteorological conditions. Studies show the *El Nino* phase of Southern Oscillations (ENSO) has also impacts on droughts in India. Crop failures, leading to acute shortage of food grains and fodder are the main effects of drought which adversely affect human life and livestock population. In order to keep off the calamity, mitigation and management of drought is the only realistic approach. The government must formulate a policy on drought management as a sub-component of National Water Policy. Drought Mitigation Centres need to be established at the state and district levels. There is a need to formulate stringent laws against misuse of ground water. Efforts should be made to streamline food subsidies and distribution measures. Other programmes include establishment of fodder banks, seed banks and micro-credit societies.

Though not much can be done to prevent natural disasters such as severe droughts, but intensity of their impact can be reduced considerably through adequate planning, early warning and implementation of action plans with the joint efforts of the government and people's involvement. Thus, there is an urgent need to utilize water resources equitably and in a sustainable manner.

## REFERENCES

Anonymous (2007). UNO Report, *Dainik Bhaskar,* A Hindi Daily, 28th August, 2007.

Bhalme, H.N. and Molley, D.A. (1980). Cyclic fluctuations in flooded areas and relationship with double (Hale) sunspot cycle. *J. Climat. & Appl. Metor.*, 22: 1041-1048.

Bose, S. and Saxena, N. (2000). Drought: The Indian Scenario. *Employment News*, 25 (9-10) : 1-2.

Chouhan, T.S. (1984). A few suggestions for future development of drought prone areas in Rajasthan. *Trans. Isdt. & Ucds.*, 9(2) : 72-79.

Das, P.K. (1984). *The Monsoons : A Perspective.* Indian National Science Academy, New Delhi.

Das, P.K. (1988). *The Monsoons*, 2nd edn. National Book Trust, New Delhi.

Jhalani, B. (2004). Drought condition in Rajasthan & their Management. *Proc. Natnl. Seminar on Challenges of Managing Drought-Prone Economy.* Dept. of Economics, Girls College Chirawa (Raj), 20-21 September, 2004.

Mahlawat, Manjeet and Singh, Mahadevi (2003). Impact of drought and its management. Chapter 3, *In: Current Environmental Issues* (Eds. B.B.S. Kapoor, Ahmed Ali, K.K. Singh and Chandrakanta). Madhu Publications Bikaner (Raj.), pp. 37-52.

Mathur, O.P. (1981). Review of Rural development programmes in arid region of Rajasthan with special references to DPAP, DDP, IRD & Antyodaya. *Trans. Isdt & Ucds.*, 6(1): 108-114.

Mathur, O.P. and Mathur, S.K. (1989). Strategies for combating drought and meeting fodder needs of livestock population with reference to soil conservation in arid and semi-arid tracts of Rajasthan State, India. *In: Proc. 6th International Soil Conservation Conference*, Addis Ababa, Ethiopia.

Oliver, J.E. and Fairbridge, R.W. (1987). *The Encyclopedia of Climatology.* van Norstrand Reinhold Co., New York.

Rai, A.N. (2004). Drought Management in India: Challenges and Strategies. *Employment News*, 29 (19) : 1-2.

Ramakrishna, Y.S., Rao, A.S. and Joshi, N.L. (1988). Adjustments to weather variations for efficient agricultural production system in arid western plains of India. *Fertilizer News*, 4: 29-34.

Ramana Rao, B.V., Ramakrishna, Y.S. and Danley, H.S. (1984a). Influence of Water availability on yield of green gram. *Mausam*, 35: 241-242.

Ramana Rao, B.V., Ramakrishna, Y.S. and Danley, H.S. (1984b). Influence of Water availability on yield of cowpea under rainfed conditions. *Annals of Arid Zone*, 23 : 63-66.

Rao, R.V.B., Sastri, A.D.R.A.S. and Ramakrishna, Y.S. (1981). An integrated scheme of drought classification as applicable to Indian arid zone. *Idojaras,* 85 : 317-322.

Rathore, M.S. (2005). *Drought Assessment and Mitigation.* IWMI, Colombo, Sri Lanka.

Sarkar, R.P. (1979). Droughts in India and their predictability. *In : Proc. Symp. Hydrol. Aspects of Droughts*. Indian Institute of Technology, New Delhi, 33-40.

Sastri, A.S.R.A.S. and Ramakrishna, Y.S. (1980). A modified scheme of drought classification, applicable to arid regions of western Rajasthan. *Annals of Arid Zone*, 19 (1) : 65-72.

Thornthwaite, G.T. (1948). An approach towards rational classification of climates. *Amer. Geography*, 37 (2) : 87-100.

# Section II

## ENVIRONMENTAL CRISIS, NATURAL RESOURCES AND CLIMATE CHANGE

# Chapter 7

# Crisis and the Environment

*Jani Bergstrom*[1&2]

## INTRODUCTION

In recent times, people have become increasingly worried about the aggravation of a variety of environmental problems. It appears that we are now better equipped to understand how vulnerable our environment is, and have even become seriously concerned about its future. On the one hand, our globe's environmental safety has been significantly influenced by a wide variety of natural phenomena. On the other hand, man is personally responsible for the self-inflicted environmental pollution, and for the aggravation of certain natural phenomena and environmental problems. The intention of this article is to clarify interaction between man and his environment in the light of a few examples, and to clearly explain a number of basic concepts and related terms. An additional objective is to awaken the reader to see the relationships between man, his environment and crises.

### Dilemma Between the Environment and Crises

What is the environment? The United Nation's Agenda 21 mentions the Human Environment as a concept that refers to "a natural and physical environment and man's relationship with this environment." The environment may also refer to the ambient culture and the social environment, in addition to the natural environment. All of these environments may be infested with crises of various types. This article examines the environment from the point of view of nature, and describes

Environmental Risk Assessment Centre, ERAC,jani.bergstrom@gtk.fi
[1]Geological Survey of Finland, Land Use and Environment, P.O.Box 1237, FI-70211 Kuopio, Finland.
[2]University of Kuopio, Faculty of Social Sciences, Department of Health Management and Economics, P.O.Box 1627, 70211 Kuopio, Finland.

the mutual interaction and impacts between man and his environment in various crisis situations.

Nature is an environment that may cause crisis situations, for man and itself alike. Nature's own development cycle and natural events (volcanic eruptions, earthquakes, etc.) may be regarded as crises in general. Alternatively, they can be seen as crises exclusively in cases where they have an impact on human beings.

Consequently, should crises be understood as situations where man is an active player or an object of events? In the light of these questions - consider the mass destructions/extinctions that have taken place on the global scale, such as the dinosaurs' disappearance at the end of the Cretaceous period some 65 million years ago, or the greatest geological mass extinction of living organisms, which took place during the Permian period some 251 million years ago. Can we say that these were crises because they did not concern people directly? Perhaps we should delimit the area in question and define crises as follows: "Unstable situations which cause a great danger or difficulty to people or the ecosystem."

Crises may be caused by the combined effect of several factors, resulting in a so-called snowball effect, or by a single major event (such as an earthquake) that may launch a massive crisis. In addition, the extent and consequences of crises vary depending on their geographical location. In Finland, for example, a jeopardy involving only 100 people may be interpreted as a crisis, whereas in China or Central America a crisis may affect the lives of millions of people. Crises of various types may include wars and other armed conflicts, environmental catastrophes (earthquakes, volcanic eruptions, etc.), disasters and human-inflicted crises (such as cultural, political and other tragedies).

**The Man-Environment Cycle**

How does the environment affect man and vice versa? We must remember that whenever man affects the environment, the environment will invariably affect him, thus creating a perpetual man-environment cycle. These relations between man and environment can be scrutinized as five different cycles.

***1. Environment - Man: The Dynamic Globe***

On the one hand, man's activity does not invariably result in environmental catastrophes. On the other hand, the environment may also constitute a major threat to people causing crises and catastrophes. These are often called as geogenic risks. A good example of this is the recent news of huge forest fires in California in late October 2007. One of the decisive factors in these bush fires was constituted by the Santa Ana winds. These so-called descending winds are generated by the Rocky Mountains due to a moist air flow passing over the mountains. Hot winds traverse a naturally dry climate feeding bush fires.

Within a short period of two days in early October 2007, the following news items were published, among others:

- Volcanic eruption kills soldiers in Yemen. 1st October 2007 (the Finnish News Agency STT)
- Heat wave in the Arctic Region. In the Canadian Arctic Region, a local research centre measured temperatures that exceeded 20 °C in the summer of 2007. These record ratings exceeded the normal level by as much as 15 degrees. 3rd October 2007 (Terra Daily, Professor Scott Lamoureux, 2007)
- Typhoon causes massive evacuation in China and Vietnam, 3rd October 2007 (the Finnish News Agency STT)

In addition to the above, earthquakes in Indonesia, for example, and their impact on millions of people were continuously in the news. Furthermore, the tsunami launched by the earthquake that occurred on Boxing Day 2004 will never be forgotten.

In late August 2007, a massive earthquake measuring about 8 on the Richter scale struck Peru, causing loss of life and property. In addition, the region, its inhabitants and the rescue teams were struck by strong aftershocks. In September 2007, the UN reported that millions of people were still suffering from floods in West and East Africa.

In the autumn of 2007 in Mexico, almost a million people had to be evacuated as a result of storms and massive floods. Cloudbursts also caused extensive landslides. For example on 6th November 2007, the news media reported that tens of people had died under a mudslide in the Chiapas State in southern Mexico.

Similar fates were reported in the autumn of 2007 as the monsoon season launched colossal damage in India and Bangladesh. A total of 2.5 million people had to leave their homes, and the floods took the lives of more than 3000. The floods also reduced the value of the crops by hundreds of millions of dollars.

Floods caused problems in other countries as well. On 8th October 2007, *Savon Sanomat* (a regional newspaper) wrote about the floods causing trouble in Vietnam. Floods also drew major attention in August 2005 as Hurricane Katrina wrecked havoc in New Orleans, leaving over 80 per cent of the city under water following the breaching of the protective embankment.

In recent times, hurricanes have caused considerable problems and destruction in the USA. The most recent hurricane, Noel (reported by STT and Reuters on 3rd November 2007) already deserves its by-name "Killer" thanks to its deadly deeds. Killer Noel wrecked havoc on the Caribbean Islands and moved north. The storm front was predicted to continue towards eastern Canada with a diminishing force. Storms, cloudbursts and floods battered the Dominican Republic, Cuba and Haiti causing significant loss of life, personal injuries, loss of infrastructure and property. These floods, cloudbursts and storms also hampered rescue work and the delivery of disaster relief assistance.

In mid-November 2007, Cyclone Sidr wrecked havoc in Bangladesh. It was feared that thousands of people had lost their lives, and about three millions were left homeless. The cyclone-induced storm surges swept over the flat coastal area of Bangladesh while cloudbursts and difficult terrain hampered rescue work and the delivery of disaster relief assistance to the victims. The loss of communication contacts and electric power prevented, for example, the immediate reporting of the number of victims and the magnitude of the devastation. For example, in Dhaka, a city with about 10 million inhabitants, most people suffered from a total blackout. According to the news, the country lost 95 per cent of its grain crops (17th November 2007, STT and Reuters).

In addition, the lethal effects of heat waves have been reported from various areas throughout the world during the summer months. The above are just a few examples of the dynamic globe's normal cycle of events. In Finland, the greatest threat is probably caused by extreme weather conditions and storms. It has been free from volcanism for millions of years, but earthquakes occur on an annual basis. These are, however, not very strong, but there was a single earthquake massive enough to collapse a church in Finland in the 1950s.

One of the best known volcanic eruptions was that of Mount St. Helens in 1980. In terms of its energy level, this explosive eruption corresponded to 27,000 Hiroshima atom bombs. Still, world history recalls much bigger eruptions, with the number of victims exceeding 90,000.

The explosive eruption that destroyed most of Santorini Island in 1366 BC had an energy level 120 times greater than that of Mount St.

Helens. As far as is known, the greatest number of victims was caused by the 1815 Tambora eruption in Indonesia where about 96,000 people lost their lives.

Mount Rainier, one of the Cascade Volcanoes, located near Seattle (USA), constitutes a future threat that will cause a problem for destruction prevention and the related rescue work planning. The volcano lies 96 kilometres from the densely populated city of Seattle with minor towns in its immediate vicinity. An explosive eruption like that of Mount St Helens would cause massive destruction to these urban areas.

The most lethal effect of volcanic activity is not the eruption as such, but the subsequent mobility of matter. When erupting, especially in an explosive manner, a volcano disgorges ash, lava and various gases. Burning pyroclastic clouds can reach speeds of 100 $kmh^{-1}$ sweeping down the volcanic flanks incinerating everything in their way, due to maximum cloud temperatures of 1000 °C. The clouds are also called *Nuee Ardente* or Tefra consisting of ash and gases (MacDonald, 1972).

Another threat factor relating to volcanic eruptions that causes destruction is the so-called Lahar slides. In practice these are mud slides consisting of pyroclastic material and water (MacDonald, 1972). Volcanoes are generally covered by snow and ice. In conjunction with an eruption, the hot pyroclastic matter melts the snow and ice layers, which are partly vaporised directly into the ambient air, while the rest converts into water causing Lahar slides. Lahar slides may be several metres in height and advance at a maximum speed of 60 $kmh^{-1}$. Ancient Pompeii, for example, was buried under slides of this type in 79 AD.

In Pompeii, Lahar slides and lava flows were not the exclusive causes of people's death. In has been found in archaeological research, with the aid of plaster casts, that some of the victims had assumed a curled up position with their mouths open as if having breathing difficulty. This proves that the majority of the victims died as a result of the ash and gases discharged by the volcano, for example. Eruptions release huge volumes of gases into the atmosphere, such as nitrogen ($N_2$), nitrogen compounds, carbon dioxide ($CO_2$), carbon monoxide (CO), sulphur dioxide ($SO_2$), sulphur (S), hydrogen sulphide ($H_2S$) and various chlorine compounds (such as HCl) [MacDonald, 1972].

The gases released in this connection have a major impact on our climate. Among other things, it has been estimated that the ash and gases released by the Tambora eruption were effective enough to have lowered the globe's average temperature by as much as 3 °C. However, the study in question has been heavily criticized, like all other studies dealing with climatic warming or cooling.

*2. Environment—Man—Environment: The Beneficiation of Natural Resources and Polluting*

Let us examine the equation between uranium and nuclear waste, for example. Uranium is made from uranite (the most common uranium ore). Uranium ore is enriched for exploitation into nuclear fuel, for example. The enrichment processes also produce nuclear waste. Unfortunately, nuclear waste emissions have been released into nature, up to catastrophic levels. During the Soviet period in Russia, for example, radioactive waste was accidentally or deliberately released into the environment from the country's nuclear fuel production plants. Over the years, in certain geographical areas, pollution reached such a level that, even today, traces of radioactive waste can be measured at hundreds of kilometers' distance from former emission sites. For example, it has been necessary to cover Lake Karachay with a concrete lid because of nuclear waste emissions which had been previously discharged into the lake. At present, the site is the most polluted place on earth.

Radioactive pollution of this type may seep through the ground into waterways and groundwater, causing a hazard in the human consumption of water at a later stage. Alternatively, waterways may dry up for a specific reason, so radioactive dust may spread by means of winds causing exposure to animals and people. It must be remembered, however, that this is not the only event in world history - several other nuclear plants have spread high volumes of radioactive waste throughout the world. After all, we still need to harness radioactive power for beneficial exploitation.

Nuclear catastrophes, the production of nuclear fuel, and radioactive waste do not constitute the only man-environment cycle for us to worry about. The infamous Aral Sea has already become a symbol of a human-inflicted environmental catastrophe.

Not only use of nuclear power, the use of other natural resources, e.g., oil has been causing problems and risks for environmental health. The whole oil refinery chain is easily vulnerable to accidents. On-shore oil fields are commonly contaminated, off-shore oil fields are huge risks for sea environment if they face an accident. Tankers and oil pipes may have leakages and, so contaminate soil or waters. Oil refineries, and the end of chain gas-stations and cars are polluting in nature. Still, we need to use this natural resource and ease our everyday lives.

### *3. Environment - Man - Environment — Man: The Effect of Natural Resources Management to Environmental Health*

Since the 1960s, the Aral Sea has shrunk in size, due to the harnessing of its tributaries and massive evaporation. Emissions and waste from the surrounding urban areas, industrial plants, agricultural areas and rivers have made the Aral Sea extremely polluted. Due to the massive evaporation, the sea's salinity level has also risen by more than 400 per cent.

In 1918, the then Russian government decided to start cotton farming in a desert area located near the Aral Sea. The project required huge volumes of water, which were extracted from the Aral Sea's tributaries. A massive irrigation network was built in the 1930s, causing a situation where the evaporation volumes surpassed those conveyed by the sea's tributaries. A few decades later, untreated waste water was piped directly into the sea destroying its stock of fish and depleting the harbour towns of their previous supply of food and fish-based industries. Sand storms and the non-arable, saline, contaminated and bare land, plus the drying of the sea and the resulting desertification, have made the local towns and residential areas almost totally uninhabitable.

In terms of an ecological catastrophe, the greatest problem is the biochemical weapons production laboratory, based on Vozrozhdeniye Island. From 1949 to 1992, pathogenic weapons were developed on the island, with anthrax being one of the most important. As a matter of fact, the island no longer exists - it has become a peninsula due to the massive recession of the Aral Sea's surface level. This means that animals can now access the area without restrictions, to spread anthrax, for example.

Recently, efforts have been made to rehabilitate the Aral Sea, for example by separating its north and south sections by means of an embankment. This has slightly improved the situation but more radical measures are still required. Therefore, the Aral Sea has been seen as a sad example of human-inflicted ecological catastrophes on a massive scale, based on mere bad decisions and the incompetent administration of the natural resources.

On the other hand, people may also cause environmental problems inadvertently, as a result of accidents, minor carcless mistakes or human error.

One of the future threats is man's interference with the globe's hydrologic cycle. Sea water is increasingly being converted into fresh water, which means that we are tampering with the natural circulation of water. Major cities like Los Angeles, Miami and Dubai must currently secure water supplies for their populations and tourists alike. This means that they have to produce millions of cubic meters of fresh water from sea water on a daily basis. Up till now, the adverse effects of this process have not been observed or investigated on a major scale. However, it is obvious that this type of interfering with the natural hydrologic cycle will cause severe problems at a later stage.

On the other hand, the massive volumes of ice on Greenland are already melting and supplying the seas with additional fresh water. The problem with this is that the Gulf Stream traversing the North Atlantic Ocean may become disrupted by the increasing fresh water supply. The Gulf Stream is generated by the oceans' thermohaline circulation where water circulation is caused and maintained by the sea water's varying salinity and temperature levels. When heading for the North Atlantic, the warm Gulf Stream passes Florida, Cuba and the Bahamas Islands and cools down when it reaches the coast of Norway. When entering the northern areas, the Gulf Stream's salinity level rises as a result of evaporation. Consequently, the Stream's water cools down, sinks deeper and turns back south again. There is a danger that the influx of additional fresh water to this sensitive salt-driven mechanism may stop the Gulf Stream. This would cause a large-scale climate change in Scandinavia and parts of Northern Europe, making the area uninhabitable due to the resulting much colder climate.

The Gulf Stream question and potential future problems have been widely discussed in media. Some scientists have arguments that the Gulf Stream is going to slow down, and the other group of scientists has been presenting measurements, showing the Stream getting stronger.

One of the future stabilizing opportunities could be the mixing of the highly saline "waste water" from the sea-water-based fresh water conversion process in a suitable proportion with the North Atlantic glaciers' melt waters. This would be used to secure and maintain the Gulf Stream's thermohaline circulation. However, we must remember that assumptions, ideas and theories of this type require extensive additional research and analyses.

## *4. Man — Environment - Man: The Effects of Anthropogenic Environmental Problems to Environmental Health*

### *4.1. The Effects of Armed Conflicts*

Wars and armed conflicts concern not only people but they may als impose considerable adverse effects on the environment. This may l illustrated through numerous examples: Huge volumes of napalm and herbicides were used during the Vietnam War, the effects of which are still seen in the local jungle areas; the oil-related damage inflicted by the Korean War, the Iraq-Kuwait War and the Lebanese Wars; plus the effects of the nerve poisons used during the Iraq War, etc. The list could be extended up until the very end of our war-related history. Unexploded bombs, ammunition containing depleted uranium, mines and aerial bombs continue to be a major threat to people, animals and the environment alike.

An additional factor to consider is the impact of military equipment, logistics, transportations and the related infrastructure on our environment. The damage caused by tanks and armored vehicles to the topsoil accelerate erosion and expose the groundwater to pollution, for example. Abandoned, leaking and rusting vehicles, barracks, prisons, mass graves, decomposing corpses, waste materials, and the like, all leave their imprints on nature and may, subsequently, cause exposure to people and animals. As a matter of fact, all areas infested with armed conflicts are similar to dumping areas without the prevention of detrimental substances into the environment. Even today people die because of abandoned chemical weapons, their soluble detrimental components and other toxic substances that have been used, for example, in the Sino-Japanese War in Manchuria in the 1930s.

The war industry produces immense volumes of substances that can have detrimental effects on man and nature. The greatest proportion of these substances and chemicals ends up in the environment, causing direct or indirect damage. Environmental vulnerability, and the use of natural resources, can be, and has been, exploited strategically in warfare. In general, the intention has been to pollute the local drinking water supplies in a sabotage fashion, which has caused long-term problems to people's lives in the infested areas after the conflict. Local areas have also been exposed to floods by exploding protective embankments so as to break the locals' resistance. Emissions from war machines of various types have also been a great burden to the environment and the climate. There have even been attempts to change the weather artificially, which is one of the most recent phenomena in environmental warfare.

In addition, warfare causes massive consumption and huge energy needs. It leads to industrial expansion, which subsequently increases the volume of air pollution and other emissions. Forests are clear cut and destroyed on a massive scale, thus accelerating erosion. Agricultural lands, grazing grounds and other means of food production are destroyed systematically.

### *4.2. Terrorism*

The new chapter for the history of risks has been recently written by terrorism. The terrorism has contributed a new element to our environmental problems and crises on a global scale. As in conventional warfare, terrorists also destroy food production areas and use biological weapons, causing environmental problems. It appears that their intention is to cause maximum damage to both the people and the environment. Terrorists hardly stop and think about environmental problems in their operations when spreading their ideology and ways, expanding their territory through chaos and fear.

### *4.3. Environmental Toxins and Chemicals*

On a global scale, thousand of hazardous chemicals are produced, transported and used, putting people and nature in jeopardy. An outstanding example of this is MTBE (methyl tertiary-butyl ether) that is used in the manufacture of petrol, for example. The compound was developed to replace the petrol-contained lead and it simultaneously improved petrol's compression tolerance and enhanced the combustion process. However, a major problem was constituted by the fact that the chemical causes contamination, even at extremely low concentration levels. A droplet of MTBE is enough to pollute a huge volume of fresh water. This may cause severe problems in nature, in groundwater areas in particular. Contamination may threaten entire aquifers in cases where the chemical seeps into groundwater. Similar compounds and chemicals may cause extensive environmental damage - making entire residential areas uninhabitable in the worst case (The Blue Ribbon Panel, 1999).

Generally speaking, environmental toxins include detrimental organic compounds, chemicals and heavy metals. Many environmental toxins are purely man-made and generated as industrial by-products, for example. These toxins are not necessarily discharged into nature on purpose, contrary to biocides, which are spread for the purpose of protecting cultivated crops, for example. In addition, agriculture and forestry use high volumes of fertilizers, which are not classified as environmental toxins. Nevertheless, these may cause a major strain on the environment, possibly resulting in health hazards, among other things. A good example of this is the wide-spread stomach complaint epidemic that occurred in the Vihti municipality in southern Finland in 2000, due to the use of liquid manure as a fertilizer.

When analyzing the conveyance of environmental toxins and detrimental substances in crisis-infested areas, for example, it is extremely important to observe whether the substance in question is easily soluble in water, fat-soluble, corrosive, volatile, or lighter or heavier than air or water. Following this, further instructions may be issued so as to more comprehensively ensure the rescue workers' safety. Substance conveyance mechanisms into and within the environment play a crucial role in crisis situations.

The most significant environmental toxins include dioxins, freons of CFC compounds, POP, PAH and PCB compounds, as well as heavy metals. A typical feature of these chemicals is that they do not decompose in nature, or their decomposition process is extremely slow.

In the 1930s, development started to find substitutes for toxic gases (such as ammonia and sulphur dioxide) that were previously used in refrigerators. Chlorofluorocarbons or CFC compounds were developed for this purpose, which are also referred to as Freons. These compounds are entirely synthetic without a natural source. CFC compounds were found to have an extremely detrimental effect on the atmosphere, especially on the protective ozone layer (located at a height of about 20-30 kilometres). The said compounds decompose at an extremely slow rate, and their chlorine content has already reached the stratosphere zone of the atmosphere (located at a height of about 15-50 kilometres). One of the consequences of this is that volcanic activity and gases, released in this conjunction, may help bring about more extensive ozone depletion by reacting with existing atmospheric compounds such as chlorine.

POP compounds (persistent organic pollutants) are extremely slow to decompose in the environment and cause serious environmental and health hazards, even at minimal concentration levels. These compounds are long-lived, and may travel long distances air-borne, for example. DDT, PCB compounds, dioxins and PAH compounds are all examples of POP compounds.

Dioxins are so-called super toxins which do not occur naturally in the environment. They are generated as industrial by-products, for example, when organic substances are incinerated in a chloridising environment. The paper bleaching process also generates dioxins, provided that chloridising is involved. Dioxins are not used as such in any process - they are purely by-products by nature. Dioxins are extremely effective poisons since a minimum lethal dose to kill a human being is only 5 mg (Freeman *et al.*, 1989).

When released into nature, dioxins will cause cell deformations and cellular malfunctions. In addition to the above, dioxins are cancer-causing carcinogens (National Research Council of the National Academies, 2006). Since the dioxins are purely synthetic, they can also be categorized as Man - Environment effect cycle (see 5 below).

PAH compounds (polycyclic aromatic hydrocarbons) are generated as a result of the incomplete combustion of organic substances, in various industries, energy production and traffic, for example. Nature may also emit the said compounds into the air, in conjunction with volcanic eruptions and forest fires. Many of these compounds have been found to cause mutations, in addition to their carcinogenic effects. In the late 18$^{th}$ century in Britain, for example, it was found that soot caused cancer in chimney sweepers. At present, road pavers and other people who handle oil-based products, among others, are exposed to the adverse effects of PAH compounds (Freeman *et al.*, 1989).

The use of PCB compounds (poly-chlorinated biphenyls) which were produced between 1920 and 1970 was stopped as they were found to cause major environmental hazards. These compounds were mainly used in transformers, varnishes and marine paints. Their environmental adverse effects were found to be problematic due to their durability, among other factors (Manahan, 1991).

Heavy metals are metals whose density levels exceed 5 gcm$^{-3}$. They are found in the bedrock, soil layers, plants and animals in nature, in the form of water-soluble ions, salts or gases, for example. Their state of aggregation is considerably affected by environmental factors (temperature and pressure). Heavy metals are released into the air, waterways and the ground as a result of man's activity and natural phenomena. Among others, heavy metals include the compounds of arsenic, cadmium, chrome (VI), copper, lead, mercury, nickel and tin. Some of these are health-promoting micronutrients in low concentrations, and their adverse effects are only detected after prolonged exposure to high concentrations. The most detrimental environmental toxins among the heavy metals are lead, cadmium and mercury. Bacterial activity, in particular, may convert mercury into methyl mercury, which is considerably more toxic than ordinary mercury.

To avoid chemical caused problems in Europe, European Union established Seveso 1 and 2 Directives, aiming at the prevention of possible catastrophes. The first directive is concerned with possible major accident hazards of certain industrial activities (The Council of the European Communities, 1982, "Seveso I") and the later one concentrates to dangerous substances (The Council of the European Union, 1996, Seveso II).

These directives are giving good guidelines for the activities in Europe. We just have to keep in mind that these Directives concern only European Union's countries and member states, not the whole globe.

*4.4. Industry and Mining*

The various industries and their raw material acquisitions also contribute to the generation of risks and crises. The mining and industrial areas may be very extensive. In the worst case, acid drainage may seep into nature from wall rock heaps or dump heaps in ore dressing sand areas, causing adverse ecological effects locally. Disasters or emissions of various types may generate significant environmental risks in the said areas. Having spread into nature, emissions and mining waste, for example, may cause problems in the ground, surface water and groundwater.

The mining and other industries are not the only causes of environmental problems, or exclusive polluters of the ground and waterways. As a result of man's activity, detrimental substances are released into the ground and waterways from numerous sources. Petrol stations, railway yards, impregnation plants, sawmills, shooting ranges, scrap yards, repair shops, vehicle depots, dumping areas, laundries and washing plants are all good examples of places where detrimental substances are used. The Finnish Environment Institute, for example, has made a national survey of thousand of locations that are suspected of being polluted. Nevertheless, it must be taken into account that some of the pollution-infested sites do not necessarily cause health-related or environmental problems. Their status may be evaluated with the aid of environmental risk analysis and related exposure risk analysis. Exposure risk analysis is geared towards assessing the potential adverse effects of detrimental substances, and their mobility, for example from the ground to living organisms (plants, animals and people): in other words, to establish their bioavailability. This means that not all environment-contained detrimental substances are bioavailable. In ecological risk assessment, if something is not bioavailable, it is not considered as an environmental risk.

***5. Man - Environment: Man Effecting to Nature***

As earlier described in 4.1 above, effects of war fall on nature, and so on to the humans. Additionally, war's effects on nature are direct. War machines are demolishing topsoil covers, bombings leave marks on nature, war infrastructure, e.g., concentration camps and barracks causes stress for environment, and different operations are normally not prioritizing environmental risks that are high in scale. War zones are rather vulnerable for erosion, since protecting vegetation has been destroyed. Unfortunate example of this we can find even these days, if we scrutinize the nature of Vietnam. Napalm and herbicides destroyed Vietnam's nature, so that those effect, even to this day, nature's normal cycle.

Quite often, we are facing environmental problems due to the poor or careless management of natural resources. Over felling and grazing has been causing erosion around the world. In addition, the lack of groundwater management has been causing drought and big problems, e.g., in parts of United States.

*5.1 Desertification*

Desertification can be regarded as a result of the combined effect of man's and the environment's activity. The clear felling of forests and rain forests, overgrazing, erosion, movements of lithospheric plates and the circulation of water are all contributing factors to desertification. The Aral Sea case was already accounted for previously in this article, but there are other similar instances which have promoted desertification over the course of time. An example is the river area of Rio Puerco in New Mexico, which has become totally eroded due to overgrazing.

The prevailing misconception is that drought is the only cause of desertification. Admittedly, drought is a great problem in certain areas but it is by no means the only cause of desertification. More often than not, the crucial factor is, however, the lack or negligent administration of the land areas in question. Here, the problematic factors are the growing population and its housing-related needs. .

## CLIMATE CHANGE

Climate warming has aroused an intensive debate in recent times. Various theories have been proposed concerning the effects of greenhouse gases, including the argument that climatic warming is a normal phenomenon, recurring from time to time on the geological time scale. It is true that the globe has undergone several glacial periods, with intervening so-called warm periods. In fact, according to a number of studies, we currently are undergoing one of the said warm periods.

In colloquial language, climatic warming generally refers to climatic change. In the course of time, the climate has changed due to natural causes without interference by man or other living organisms. Nonetheless, human activity has influenced climate change notably. Thereby, Intergovernmental Panel on Climate Change defines climate change as follows "any change in climate over time, whether due to natural variability or as a result of human activity" (IPCC 2007: 2). The said variation has been influenced by the movements of the lithospheric plates (continental plates) which have subsequently affected the flow of ocean currents and the oceans' thermal mechanisms in due course. Consequently, this has had a major impact on our climate. Other factors having a significant influence on our climate include volcanism, the Earth's orbit around the sun, its axial tilt and the related precession (movements of the spinning-top type), and its nutation or "wobble". In addition, major asteroid collisions are believed to have caused extremely significant effects to the globe's climate over the course of time. According to some theories, a number of extinctions were caused by a climate change resulting from an asteroid collision.

At present, the effect of the sun's activity on climatic warming is being intensively debated. Among others, Live Science states on its website, in a letter mailed on 12th March, that other planets have experienced climatic warming simultaneously with Earth. This has been regarded as noteworthy evidence of the sun's activity on climatic warming, in addition to other geological, geochemical and isotopic measurement results. It has been found that the histories of these two separate factors intercorrelate extremely well.

On the one hand, man may play a significant role in certain natural phenomena, such as the contribution of air pollution to the greenhouse effect and in subsequent climatic warming. On the other hand, this influences the occurrence of hurricanes, and their intensity levels, in the Americas, for example. An additional major factor is constituted by the felling of rain-forests, the so-called destruction of the Earth's lungs. This is prone to increase the Earth's carbon dioxide levels. The air pollution caused by the various industries and energy production is another significant anthropogenetic factor.

The effect of man's activity on climatic warming may well be of a dual nature. On the one hand, air pollution prevents the sun's thermal radiation from escaping back into space, while, on the other hand a significant portion of it can never penetrate the pollution-generated barrier surrounding the globe. Nevertheless, it must be borne in mind that climatic warming is very likely to be the combined result of several factors.

## ENVIRONMENTAL SCARCITY

One huge future-related problem will be environmental scarcity. The population growth and our decreasing living space will cause severe problems in future since the continual loss of existing grazing ground areas, fresh water supplies and other resources will generate an increasing number of armed conflicts. Thomas F. Homer-Dixon has approached the issue by means of the Environmental Scarcity concept.

According to him, armed conflicts are caused by the depletion of natural resources, their pollution and unequal distribution between the various population groups. This results in the generation of new conflicts which will aggravate the scarcity of natural resources causing increasing poverty and growing political unrest. The outcome is a vicious circle of conflicts.

Homer-Dixon defines the living space scarcity and its consequences as per the following three hypotheses (Laakkonen and Vuorisalo, 2007):

1. The diminishing supply of fresh water and the shrinking agricultural land and grazing ground areas cause a scarcity of basic commodities (simple-scarcity), which generates conflicts and so-called resource wars.
2. The first hypothesis leads to migration-generated group identity conflicts between the so-called environmental refugees and the existing local population (ethnic minorities and cultural problems, social and socio-economic problems, poverty, rebellions, etc.)
3. These will result in increasing economic deprivation, causing civil unrest, uprisings and poverty-generated wars.

Due to our continually growing population we must plan increasingly larger urban and residential areas. People will have to live closer to geohazardous areas due to the lack of space in urban areas, the extent of uninhabitable areas, geographical distances and the lack of communication connections. When planning new infrastructures and residential areas, we must increasingly observe natural and environmental risks - what man can do to the environment so as to avoid the scarcity of natural resources - and what the environment can do to man.

These belong to the phenomena that we cannot prevent, but can alleviate their effects through precaution and analysis, or perhaps the best case scenario is to circumvent major catastrophes in advance.

It must be borne in mind, however, that the scarcity of living space no longer concerns people alone. An alarming example of this was reported by the AFP news agency on 22nd October 2007: macaque monkeys attacked the Mangliawan village in Indonesia's East Java province. People had destroyed their natural habitat which led to the macaques' violent behavior towards the people. This kind of animal behavior has been experienced, for example, in South-East Asia rather often.

## CONCLUSIONS AND FUTURE PROSPECTS

Finland has become increasingly aware of the environment and crises. The two issues are closely intertwined. In addition, efforts have been made to export the related Finnish know-how. Various crisis management operations and disaster relief assistance projects have been in existence for a long time. A new element in these efforts has been constituted by awareness of the environment and environment-induced problems, the advance prevention of pollution, and the management and assessment of environmental risks.

The Environmental Risk Assessment Centre (ERAC) and the Crisis Management Centre Finland (CMC Finland) have started activities in Kuopio and developed environmental risk assessment methods in view of crisis management needs and environmental safety (the iCrisis theme). This constitutes the frame of reference for the use of environmental expert networks and the related methodology in crises. The expert method is based on a pre-crisis risk analysis and uses the iCrisis system (Intelligent Method for Decision Making in Crisis Management). The system exploits a modeling procedure that supports the processing of risk-related information, assessment of the magnitude of risks involved plus that of the action to be taken, and the decision-making process. The objective of this development project is to refine the expert method and iCrisis system so as to comply with the requirements of local, regional, national and international crises. In addition, an expert support network will be built during the project for decision-makers working in crisis-infested areas, to facilitate their work. The expert method still requires additional development. The ready-made method will provide a trailblazing operating model for crisis management purposes. At best, it will help users predict and even prevent imminent crises. After all, our common goal is to save people and ensure a better and safer future for people and nature alike.

## REFERENCES

Chadha, R.K., Papadopoulos, G.A., Karanci, A.N. (2006). *Disasters due to natural hazards*. Published online: 17 November 2006, Springer Science and Business Media B.V.

Chandong Chang and Hyuck-Jin Park (Guest Editors). (2007). *Proceedings of the 6th Asian Regional Conference on Geohazards in Engineering*

*Geology*. The Korean Society in collaboration with IAEG (International Association for Engineering Geology and the Environment) National Groups (ICRF-2007-037-C00023).

Church, R., Quinn, S., and Salmon, S., The Scotland and Northern Ireland Forum for Environmental Research (SNIFFER)/t2003). Final Report, 6/7 Newton Terrace, Glasgow, G3 7PJ, Scotland, p. 52.

Eronen, M., Saarnisto, M., and Salonen, V.P. (2002). Kaytannon maaperageologia. Otava Kirjapaino Oy, Otabind, Turku p. 237.

Fetter, C.W. (2001). *Applied Hydrogeology*. International Edition, the Fourth Edition. Pearson Education International, New Jersey, USA.

Freeman, Harry M.(1989). *Standard Handbook of Hazardous Waste: Treatment and Disposal*. McGraw Hill, New York.

Ganoulis, J.G. (1994). *Engineering Risk Analysis of Water Pollution, Probabilities and Fuzzy Sets*. Federal Republic of Germany, Weinheim, Germany, p. 306.

German Advisory Council on Global Change (Eds.) (1999). *Strategies for Managing Global Environmental Risks: Annual report*. Executive Summary, Heidelberg.

Grathwohl, P., Halm, D., Bonilla, A., Broholm, M., Burganos, V., Christophersen, M., Comans, R., Gaganis, P., Gorostiza, I., Hohener, P., Kjeldsen, P., and van der Sloot, H. (2003). *Guidelines for Groundwater Risk Assessment at Contaminated Sites (GRACOS)*. Campus Druck. Tubingen. Germany, p. 66.

Harris, R.H., Highland, J.H., and Humphreys, K. (1984). Comparative Risk Assessment: Tolls for Remedial Action Planning. *Hazardous Waste*, 1(1): 19-33.

Homer-Dixon, Thomas F. (1994). *International Security*, 19(1):5-40, (Summer 1994).

Homer-Dixon, Thomas F. (1994). Environmental Scarcity and Violence Conflict. *International Security*, 19(1): 76-116.

Homer-Dixon, Thomas F. (1999). *Environment, Scarcity and Violence*. Princeton University Press, Princeton, N.J.

Intergovernmental Panel on Climate Change (IPCC) (2007a). *Climate Change 2007: The Physical Science Basis*. Intergovernmental Panel on Climate Change, Cambridge University Press, Cambridge.

Intergovernmental Panel on Climate Change (IPCC) (2007b). *Climate Change 2007: Impacts, Adaptation and Vulnerability*.

Intergovernmental Panel on Climate Change, Cambridge University Press, Cambridge.

Intergovernmental Panel on Climate Change (IPCC) (2007c). Climate Change 2007: *Mitigation to Climate Change - Summary for Policy-makers.* Paris. Download: http://www.ipcc.ch/SPM040507.pdf, 29.05.2007.

Jones, K.C. (1991). *Organic Contaminants in the Environment.* Environmental Management Series, Elsevier Scicnce Publishers Ltd., Crownhouse, Essex, England.

Jorgensen, P., Sorensen, R., and Haldorsen, S. (1995). *Kvartaergeologi. Norges Landbrulcsh£gskjoje, Institutt for jord- og vannfag, Landbruksforlaget*, p. 344.

Kreye, R., Ronneseth, K., and Wei, M. (1998). Ministry of Environment, Lands and Parks, Water Management Division, Hydrology Branch Province of British Columbia, Webpage: http://wlapwww.gov.bc.ca/wat/aquifers/Aq_Classification/Aq_Class.html#04.

Laakkonen, Simo, Vuorisalo, Timo (2007). *Sodan ekologia.* Tammer-Puino Oy, Tampere.

Lamoureux, Scott (2007). Arctic Heat Wave Stuns Climate Change Researchers. *Terra Daily,* webpage: http://www.terradaily.com/reports/ Arctic_Heat_Wave_Stuns_Climate_Change_Researchers _999.html.

Leeson, J., Edwards, A., Smith, J.W.N., and Potter, H.A.B.(2003). *Environment Agency: Hydrogeological Risk Assessments for Landfills and the Derivation of Groundwater Control and Trigger Levels.* Environment Agency, Rio House, Waterside Drive, Aztec West Almondsbury, Bristo, p. 100.

Lilly, A., Edwards, A.C., and McMaster, M. (2003). Microbiological Risk Assessment Source Protection for Private Water Supplies: Validation Study. Report prepared for the Scottish Executive by the Macaulay Institute. The Macauley Institute, p. 53.

MacDonald, Gordon A. (1972). *Volcanoes.* University of Hawaii, Prentice Hall, INC. Eaglewood Cliff, New Jersey.

Manahan, Stanley E. (1991). *Environmental Chemistry*, 5th Edition, Lewis Publishers, INC. Chelsea, Michigan, USA.

Madsen, H.B., Pedersen, S.E., and Rask, N. (2003). Odense Pilot River Basin, Provisional Article 5 Report pursuant to the Water Framework Directive. Fyn County, Nature Management and Water Environment Division, Environmental and Land Use Management Division. MOURET AD & DESIGN, Denmark, p. 132.

Malkki, E. (1999). *Pqhjavesi ja pohjaveden ymparisto.* Kustannusosakeyhtio Tammi, Helsinki. Tammer paino Oy, Tampere, p. 304.

Marttunen, M., and Hellsten, S. (2003). *Heavily Modified Waters in Europe: A Case Study of Lake Kemijarvi*, Finland, Helsinki. The Finnish Environment Institute. Edita Prima Ltd., Helsinki, p. 60.

Mediaviestimet mm. Savon Sanomat, televisio, lltalehti, Helsingin Sanomat, Internet.

Meteorological Institute of Finland. (2007). webpage, http://www.fmi.fi.

Meteorological Institute of Norway, (2005). webpage, http://met.no

National Research Council of the National Academies |(2006). *Health Risks from Dioxin and Related Compounds Evaluation of the EPA Reassessment.* Committee on EPA's Exposure and Human Health Reassessment of TCDD and Related Compounds, Board on Environmental Studies and Toxicology Division on Earth and Life Studies. The National Academies Press, Washington, DC.

North Carolina States University (2004). College of Agriculture & Life Sciences, webpage. http://www.soil.ncsu.edu.

Ojala, Antti, Espoo (2007). *Jaakausiajan muuttuva ilmasto ja ymparisto.* GEOLOGIAN TUTKIMUSKESKUS, Opas 52.

Palmer, R.C., and Lewis, M.A. (1998). Assessment of Groundwater Vulnerability in England and Wales. *In: Groundwater Pollution, Aquifer Recharge and Vulnerability* (Ed., N.S. Robins). Geological Society, London, Special Publications, 130, 191-198.

Robson, Mark, and Toscano, William (2007). *Risk Assessment for Environmental Health.* First Edition, Published by Jossey-Bass, A Wiley Imprint, San Francisco, United States.

Schneiderbauer, Stefan (2007). *Risk and Vulnerability to Natural Disasters-from Broad View to Focused Perspective.* Freie Universitat Berlin, Naturwissenschaftliche Fackultat Fachbereich Geowissenschaften.

Schwartz, Peter and Randall, Doug (2003). *An Abrupt Climate Change, Scenario and its Implications for United States National Security.* October, Published by the Pentagon.

Scottish Environment Protection Agency (SEPA)(2004).webpage: www.sepa.org.uk

SNIFFER, Scotland and Northern Ireland Forum for Environmental Research, (2004). webpage: http://www.sniffer.org.uk/.

Suomen Suurlahetyston tiedote Lima, Perukeskiviikko 22.8.2007, klo. 13.15 (UTC-5) http://www.finlandiaperu.org.pe/su-noticia7.htm.

Than, Ker (2007). Sun Blamed for Warming of Earth and Other Worlds, *Live Science*, webpage, posted: 12 March 2007, 07:27.

http://www.livescience.com/environment/070312_solarsys_warming.html

The Blue Ribbon Panel (1999). *Achieving Clean Air and Clean Water: The Report of the Blue Ribbon Panel on Oxygenates in Gasoline.* September 15, 1999, Environmental Protection Agency, http/7 www.epa.gov/otaq/consumer/fuels/oxypanel/r99021.pdf

The Council of the European Communities (1982). Original Seveso Directive 82/501/EEC ("Seveso 1"), Council Directive of 24 June 1982 on the Major-Accident Hazard of Certain Industrial Activities (82/501/ EEC), Brussels.

The Council of the European Union (1996). Seveso II Directive 96/82/EC, Council Directive Council Directive 96/82/EC of 9 December 1996. On the Control of Major-Accident Hazards Involving Dangerous Substances, Brussels.

United Nations Environmental Programme, International Programme on Chemical Safety (2004). *JPCS Risk Assessment Terminology,* Part 1 and 2, World Health Organization Geneva.

US Global Change Research Program (2003). webpage: www.USGCRP.gov.

Waterways Experiment Station Corps of Engineering, U.S. Army (1951). *Time Lag and Soil Permeability in Ground-water Observation.* Army - MRC Vicksburg, Mississippi, 1951. Bulletin No. 36. p. 50.

Vik, E.A., Breedveld, G., and Farestveit, T. (1999). *Guidelines on Risk Assessment of Contaminated Sites.* Statens Forurensninestilsvn (SFT), Veiledning 99:.01A TA- nummer 1629/99, p. 103.

Woldt, W., Dahab, M., Bogardi, I., and Dou, C. (1996). Management of Diffuse Pollution in Groundwater under Imprecise Conditions using Fuzzy Models. *Wat. Sci. Tech.*, 33(4-5): 249-257.

Worrall, F., and Kolpin, D.W. (2003). Direct Assessment of Groundwater Vulnerability from Single Observations of Multiple Contaminants. *Water Resource Research*, 39 (12): 1345.

## Chapter 8

# Natural Resources and Environmental Management for Sustainable Development

*K.K. Singh[1], Ali Akram[2], Gayatri Verma[3], Vijay Pal Singh[4] and Mahadevi Singh[5]*

## INTRODUCTION

The prefix "Sustainable" is now invariably used. The natural resources of land, water, forests, fisheries, etc. constitute the basic support systems for life on earth. Directly support the livelihood of hundreds of millions of people, particularly the rural poor, and contribute to agricultural and economic growth. The notion of sustainability; or Sustainable development is currently being discussed as a focal theme of development, planning and other associated aspects. In the wake of self - defeating current mode of development and recurrent natural calamities people have to think over the faults, shortcomings, weaknesses, discrepancies and limitations of the ongoing development process and production system (Swaminathan, 1996).

Human civilization has passed through various stages of development and different revolutions, With each of these, the forms and modes of man's interaction with his environment has been constantly changing at an alarming rate. It was only after the industrial revolution followed by the transport revolution and consequent urbanization that led to degradation of environment. There is an apprehension of an environmental catastrophe in near future.

[1]Project Directorate (Research), Agriculture & Soil Survey, Krishi Bhawan, Bikaner (Raj.) - 334001, India.
[2]Office of the DPC, SSA, Bikaner (Raj.) - 334001, India.
[3]Department of Chemistry, BSA College, Mathura (UP) - 21004, India.
[4]Department of Agricultural Zoology & Entomology, RBS College, Bichpuri, Agra (UP) - 283105, India.
[5]Sophia Sr. Secondary School, Bikaner (Raj.) - 334001, India.

In the past industrial revolution era the world has witnessed progress in all walks of life. The decades of the 1960s and 1970s have been marked by the intensification and spread of pollution. But the 1980s have been a reorientation of some environmental thinking. In 1980s the US 'Global 2000 Report' appeared to confirm environmentalist prophesies about the consequences of the neglect of the global common interest and the over - exploitation of the open access resources. Although the term Sustainable development came into vogue sometimes during 1970's, it became more popular in the last two decades, especially after World Conservation Strategy and Brundtland Report. The rejection of the physical limits of growth thesis, the appropriate role of market forces in the development process, the role of poverty in natural resources degradation and the need to recognize and build on common interests were accepted. It is being realized that after several decades of planned development the plight of the poor is still worse, the number of malnourished and illiterate people is larger than ever before; the violence against nature is such that our very survival is threatened, more and more communities are becoming vulnerable; women continue to be marginalized and disempowerment. It is also being felt that ongoing development process is carrying certain gross theoretic and practical fallacies which are contradictory, self-defeating, anti-masses, anti-ecological and so on. A fundamental problem that has resulted from the imbalance of our values, is our obsession with unlimited growth. Never - ending economic growth syndrome is accepted as a dogma literally by all economists and politicians, who out of sheer naivety, assume that it is the only way to ensure that material wealth will percolate down to the poor. In fact, development has to be within the existing, cultural, social and economic milieu of the country.

The term sustainable development was first used at the time of *Cocoyoc Declaration* on 'Environment and Development' in the early 1970s, meant for attaining eco-friendly development. Sustainable development aims for promoting growth, alleviating poverty, and protecting the environment that are mutually supportive objectives in the long-run but not in the short-run (World Bank, 1987). Moreover, the idea of sustainable development (SD) is not new to India, Gandhiji was one of the advocates of SD and his philosophy was based on 'Simple Living and High Thinking'. There is a growing awareness among the people about the issues of development, ecology and health.

Therefore, search of alternative development processes is being urgently felt. Sustainability is a key concept in the alternative economic development paradigms that have been advocated. The present article explains the conservation and management of natural resources for sustainable development.

## THE SUSTAINABLE DEVELOPMENT: CONCEPT, SCOPE AND DEFINITIONS CONCEPT AND DEFINITIONS

The term sustainability has its origin from the Latin word *sustinere* that means to 'hold up', 'to endure'. The Oxford Reference Dictionary refers sustain as 'endure without giving way', 'to support especially for a long period'. The meaning given by Collins Dictionary is 'to maintain', 'to support'. Similarly, Webster's New Collegiate Dictionary describes sustainability as 'to give support', 'to keep up', 'prolong'. From these, one can infer that the general meaning of sustainability is to maintain or to support a programme /project for a long period that is not possible without effective and efficient people's participation.

There is an enormous and diverse flow of literature in recent years concerned with the concept of sustainable development (SD), particularly after the Rio Earth Summit in 1992. The term 'sustainable development' was brought into common use by the World Commission on Environment and Development (Bundtland Commission) in its seminal 1987 report, *"Our Common Future"*. The idea of sustaining the earth has proved a powerful metaphor in raising public awareness and focuses on the 'need for better environment stewardship'. Many definitions have been suggested and debated thereby expressing a range of approaches linked to different world views (Turner, 1993). The proliferation of the literature on the concept of sustainable development followed its popularization by the World Commission on Environment and Development. WCED (1987) also known as Brundtland Report -where it played a central role in defining the environmental aspects of economic development. The report defines sustainable development as "development that meets the needs of the present generations", is strongly endorsed by the Human Development Report (1991). A process of change in which exploitation of resources, the direction of investments, the orientation of technology development, and institutional change are in harmony and enhance both current and future potential to meet human needs and aspirations.

This definition of WCED rightly makes the long-term realization of human needs and aspirations central to the concept of sustainable development (SD). Both in equity dimension (intra-generational) and (inter-generational) and a social/psychological dimension are clearly highlighted by this definition. If a society accepts the desirability of the goal of SD, it must development economically and socially in such a way that it minimizes the effects of its activities, the costs of which are borne by future generations.

The Brundtland Commission also highlighted the needs of the world's poor, to which urgent attention should be given and the idea of limitations imposed by the state of technology and several organizations on the society's ability to meet the needs of the present and future. The central rationale for SD is, therefore, to increase people's standard of living and, in particular, the well being of the least advantaged and deprived sections of society, while at the same time, avoiding uncompensated future costs (Turner, 1993). The definition of SD in WCED, 1987 report, highlights the important issue of the sustainability only. SD is also often defined as development that improves health care, education and social well being. Such human development is now recognized as critical to economic development (UNDP Report, 1991) and timely stabilization of population (WCED Report, 1987).

Sustainable development is defined by UNDP in its Human Development Report (1991) as "development that improves health care, education and social well being". Such human development is now recognized as critical to economic development and to early stabilization of population. It further states that "men, women and children must be the center of attention with development woven around people and not people around development."

Goodland and Ledac (1987) stress that "using renewable natural resources in a manner that does not eliminate them or otherwise diminish their 'renewable usefulness' for future generation as unsustainable development".

The World Conservation Strategy (1980) defined sustainability as "the management of the human use of the biosphere so that it may yield the greatest sustainable development to present generation while maintaining the potential to meet the needs and aspirations of future generations."

World Conservation Union (1991) defines sustainable development as "improving the quality of human life while living within the carrying capacity of supporting ecosystem." The Report focuses on sustainable development as a process requiring simultaneous global progress in a variety or dimensions; economic, human, environmental and technological.

"Sustainable development is development that improves the long term health of human and ecological systems" (Wheeler, 2004).

"Sustainable development describes a process in which natural resource base is not allowed to deteriorate. It emphasizes the hitherto unappreciated role of environmental equality and environmental inputs in the process of raising real income and the quality of life" (Pearce and Warford, 1993).

The Food and Agricultural Organization (FAO, 1995) defines Sustainable development as "The management and the conservation of the resource base and the orientation of technological and institutional changes in such a manner as to ensure the attainment and continued satisfaction of human needs for present and future generations. Such sustainable development is environmentally non-degrading, technically appropriate, economically viable and socially acceptable (cited in Reddy, 1995;A-21).

Though sustainable development is the acknowledged subject of most recent development thinking, little head way appears to have been made in terms of rigorous definition of the concept. Therefore, not surprisingly, very little efforts have been made to operational sustainable development and to show how it can be integrated into practical decision making. The use of the term development rather than economic growth implies acceptance of the limitations in the use of indexes such as Gross National Product (GNP) to measure the well being of nations. Now-a-days, development embraces wider concerns, e.g., the quality of life, educational attainment, nutritional status, and access to basic freedom and spiritual welfare.

The emphasis on sustainability suggests that what is needed, is a policy effort aimed at making these development achievements last longer into the future. According to Winpenny (1991) sustainable development is that which leaves our total patrimony, including natural environmental assets, intact over a particular period. We should be able to bequeath to future generations the same capital, embodying opportunities for potential welfare that we currently enjoy. Perhaps the best definition of sustainable development is given by Repetto (1998). For him, sustainable development is a development strategy that manages all assets, natural resources, as well as financial and physical assets, for increasing long-term wealth and well being. Sustainable development as a goal rejects policies and practices that support current living standards by depleting the production base, including natural resources and leaves future generations with poorer prospects and greater risks than our own (Mohanty, 1999).

The definitions of sustainable development, therefore, are many, depending on the nature of the problem addressed (Arnold, 1989). The sustainable development as applied to the Third World is the increasing of material standard living of the poor at the 'grass roots' level and its primary objective in general is "reducing the absolute poverty of the world's poor through providing lasting and secure livelihoods that minimize resources depletion, environmental degradation, cultural disruption and social unstability" (Barbier, 1987). To achieve an all round development goal and strengthen the roots of development further, one has to dwell upon natural resource management.

Some scholars have expanded the definition of SD still further to include a rapid transformation of the technological base of industrial civilisation (Speth, 1989). They point out that new technology is needed that is cleaner, more efficient and more sparing of natural resources in order to reduce pollution, help stabilize climate and accommodate growth in populations and economic activity (Heaton *et al.*, 1991). In economic terms, SD is defined as "economic system in which the number of people and the quantity of goods are maintained at some constant level that is ecologically sustainable over time and meets at least the basic needs of all members of the population (Khoshoo and Sharma, 1992).

Increasingly, definitions of SD attempt to cut cross or encompass several aspects or dimensions. The new strategy outlined by the World Conservation Union -Caring for the Earth - defines SD as "improving the quality of human life while living within the carrying capacity of supporting ecosystems (IUCN, 1991). This report focuses on SD as a process requiring simultaneous global progress in a variety of human dimensions ; economic, human, environmental and technological. Thus, the strategy to ensure SD has at least four important elements ; it has to be ecologically harmonious and economically efficient, and must aim at local self-reliance and offer equity with social justice. It is, thus, clear that SD is a multi-dimensional concept.

Sustainable development is defined as a pattern of social and structural economic transformations, i.e., development which optimizes the economic and societal benefits available in the present, without jeopardizing the likely potential for similar benefits in the future. A primary goal of sustainable development is to achieve a reasonable and equitably distributed level of economic well-being that can be perpetuated continually for many human generations.

Sustainable development implies using renewable natural resources in a manner which does not eliminate or degrade them, or implies using non-renewable (exhaustible) mineral resources in a manner which does not unnecessarily prelude easy access to them by future generations. Sustainable development also requires depleting non-renewable energy resources at a slow enough rate so as to ensure the high probability of a orderly society transaction to renewable energy sources.

Based on similar arguments, sustainable development has been alternatively defined in various manners also, some of them are as follows:

While many definitions of term have been introduced over the years, the most commonly cited definition comes from Brundtland Commission Report (1987), *Our Common Future,* which states that sustainable development is development that "meets the needs of the present without compromising the ability of future generations to meet their own needs". Therefore, the development plans have to ensure :

- Sustainable and equitable use of resources for meeting the needs of the present and future generations without causing damage to environment;
- To prevent future damage to our life-support systems;
- To conserve and nurture the biological diversity, gene pool and other resources for long-term food security.

—State of the Environment Report - 1999,
Ministry of Environment of Forests,
Govt. of India.

**Sustainable Development Portal**

Sustainable development does not focus solely on environmental issues. It can be conceptually divided into four general dimensions : social, economic, environmental and institutional. The first three dimensions address key principles of sustainability, while the last one address key institutional policy and capacity issues. In support of this, several United Nations texts most recently the 2005 World Summit Outcome Document refer to the "interdependent and mutually reinforcing pillars" of sustainable development as economic development, social development, and environmental protection (Fig. 1).

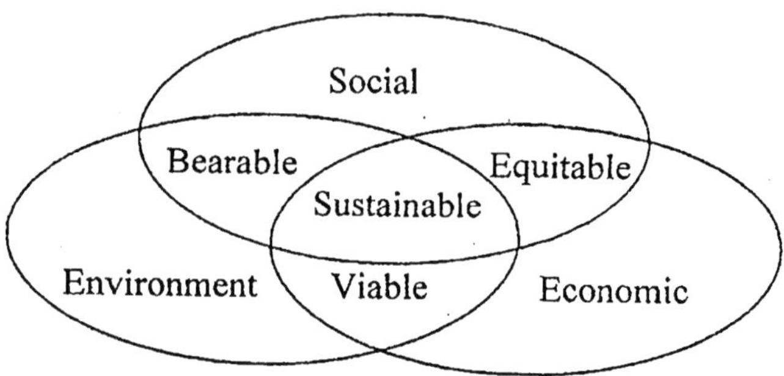


**Fig. 1 : The Sustainable development Portal**

The Universal Declaration on Cultural Diversity (UNESCO, 2001) elaborates further the concept by stating that "cultural diversity is as necessary for humankind as biodiversity is for nature"; it becomes "one of the roots of development understood not simply in terms of economic growth, but also as a means to achieve a more satisfactory intellectual, emotional, moral and spiritual existence." In this vision, cultural diversity is the fourth policy area of Sustainable development.

Some research activities begin from this definition to show that the environment we inherited and that we will transmit to future generations is a combination of nature and culture. The network of Excellence "Sustainable development is a diverse word." It integrates multi-disciplinary capacities and interprets cultural diversity as a key element of a new strategy for sustainable development. The United Nation Division for Sustainable Development lists the following areas as coming within the scope of sustainable development. These are : agriculture, atmosphere, biodiversity, biotechnology, capacity building, climate change, consumption and production patterns, demographics, desertification and drought, disaster reduction and management, education and awareness, energy, finance, forests, fresh water, health, human settlements, indicators, industry, information for decision making and participation, integrated decision making, international law, international co-operation enabling environment, institutional arrangements, land management, major groups, mountains, development strategies, oceans and seas, poverty, sanitation, science and technology, small islands, sustainable tourism, toxic chemicals, trade and environment, transport hazardous radioactive and solid wastes, and water.

Sustainable development is a notoriously ambiguous concept, because a wide array of views have fallen under its umbrella. The concept has included nations of weak sustainability, strong sustainability and deep ecology. Different conceptions also reveal a strong tension between eco-centrism and anthropocentrism. Thus, the concept remains weakly defined and widely debatable and criticised as a precise definition.

In the third world, improved living levels depend largely on increased consumption of resources. Therefore, this definition of sustainable development means that present levels and methods of resource availability in the future will decline. Sustainable development is a development process that not only generates economic growth but distributes its benefits equitably, that regenerates the environment rather than destroying it. It is intrinsically inexact concept, which can not be measured but can be a general guide to policies that have to be with investment, conservation and resource use. In short, sustainability is an injunction not to satisfy ourselves so that we have to the future the option or the capacity to be as well off as we are. To make sustainability, the advantage of the principle of sustainability can be taken. A desirable sustainable system have five R's, i.e., resistance, resilience, regeneration, redesign and replenishment. At the same time it must be able to resist, absorb, recover, adjust or be restored and simultaneously should be capable enough to withstand the periodic ecologic and economic adversity (Kohli *et al.,* 1997).

In practice, however, SD means different things for different nations. All countries are different and in particular there are profound differences in the conditions of life and in outlook between the rich - industrialized countries, and the poor - primarily rural countries of the world. Broadly speaking, developed countries lay emphasis on economic and technological aspects. The general feeling is that the current development processes can continue, provided the technological innovations are repaid and appropriate enough to reduce environmental ill effects. In contrast, the developing countries present totally different perspectives. In their view, poverty is the greatest pollution of environment and hence they must accelerate economic growth to meet the basic needs of the people. Alleviating absolute poverty has important practical consequences for SD, since there are close links between poverty, environmental degradation and rapid population growth. Eradicating poverty on a massive scale requires accelerated pace of development, which may lead to environmental crisis. They also believe that the present environmental crisis has mainly been the creation of rich advanced countries, and hence they have a responsibly to assist the latter with finance and latest environment friendly technological know-how to promote economic growth. The perspective had been a major thrust of deliberations in the 1992 UNCED Rio Conference on Environment.

On per capita basis inhabitants of industrial countries use many times more of the world's natural resources than do inhabitants of developing countries. Consumption of energy from fossil fuels, for example, is 33 times higher in the United States than in India and 10 times higher in countries of the Organization for Economic Cooperation and Development (OECD), on an average than in the developing countries (WRI, 1992). Therefore, industrial countries have a special responsibility for leadership in SD. Rich countries have the financial, technical and human resources to take the lead in developing cleaner, less resources - intensive technologies, in transforming their economies to protect and work with natural systems, and in providing more equitable access to economic opportunities and social services within their societies. The leadership also means providing - as an investment in the future of the planet of - technical and financial resources to support SD in other countries.

Thus, sustainable development is not a static concept and can not be defined once for all. Making the concept of sustainability precise, however, has proved to be difficult. It is not plausible to argue that all natural resources should be preserved. Successful development will inevitably involve some amount of land clearing, oil drilling, river damming, and swamp draining. The concept goes beyond the parameter of self-reliance and self-sufficiency though the two are often attained together. Sufficiency does not necessarily always ensure sustainability. In short, sustainable development calls for multifaceted and comprehensive interventions to address wide varieties of problems, simultaneously using different types of policy approaches, each designed specifically with reference to a particular issue. Resilience is the watch-word of this concept. It is a dynamic process and is applied by different countries of the world in tune with their own cultural, political and economic perspectives.

**Sustainability : A Perspective**

The imperatives of economic efficiency, ecological health and socio-political democracy are interdependent. Sustainability is :

(a) A normative standard or social goal, and

(b) A vision of the future with many origins.

There are at least two perspectives on sustainability, namely,

**(a) The Ecological View**

According to this view, sustainability is promised on the integrity of the ecosystem because there is no humanmade substitute.

**(b) The Social Science View**

According to this view, sustainability also require democratic process and a vital economy.

**Development Choices and Challenges**

Though the paradigm of sustainable development is riddled with contradictions and conflicts at the conceptual as well as operational level, the much - propagated conflict between economic development and environmental concerns seems to be less in the light of socio-ecological attributes like awareness and attitudes. How attitudes and lack of awareness affect and obstruct the processes of planned change and economic growth is neatly documented in the writings of all development practitioners, notably sociologist and anthropologists. These characteristics are more crucial in the context of agricultural sustainability in India, which is known for its inter-regional, intra-regional and cultural variation. Lack of comprehensive or integral understanding of there aspects both at the conceptual and operational levels makes the question of agricultural sustainability ambiguous. In this context, it can be argued that the models of sustainable development in agriculture, which focus on the people as their primary concern, should not stop at providing them with livelihood alone (Mohanty, 1999).

**The Currant Trends in the Ecology of an Industrial Planet**

In the life of the earth, 200 years is a more flicker of time. Yet within the past two centuries the rise of industrialism has transformed the planet in ways that natural processes and previous civilizations would have taken millennia to achieve. In this short era of 'modernity' we have wrought dramatic changes to the environment, the most far - reaching being our effect on the chemistry of the atmosphere and the genetic diversity of the planet. These changes have given rise to fear of global environmental crisis, and to calls for a shift from exploitative industrialism

- 'business - as - usual' - to something called 'sustainable' development (Swanson and Barbier, 2005).

The key issue for sustainable development is the magnitude of the changes induced by the socio-economic parameters. There is a 'techno centric' school of thought which suggests that the negative consequences of these trends can be overcome or managed; and that human technological prowess will allow indefinite economic growth, will help to manage and eventually contain global population increase, and will deliver ever higher living standards (Fig. 2) (Simon and Khan, 1985).

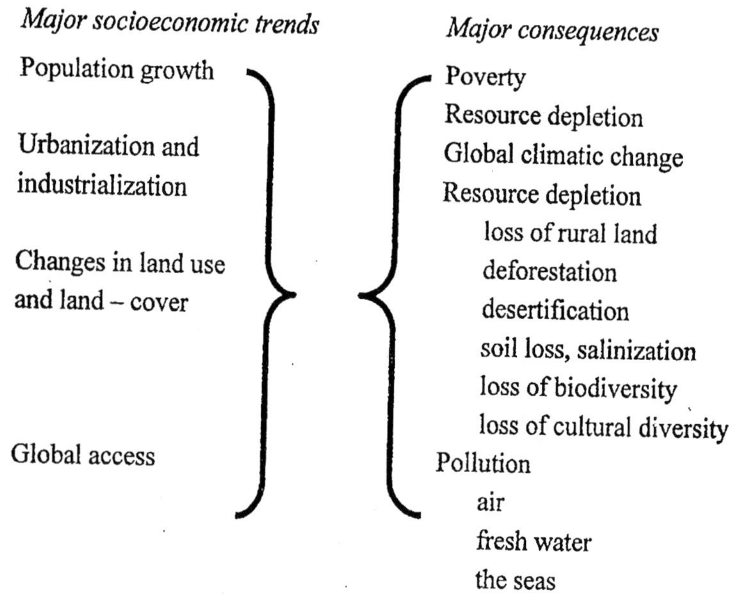


**Fig. 2 : Major world socio-economic trends and sustainable development**

However, this perspective fails to take sufficient amount of the delicate balance of complex ecosystems and the possibility of dynamic negative changes being triggered by excessive human growth. For example:

- The current world population growth of some 80 million people per year - although down from the peak of 87 million in 1990 - virtually ensures poverty, undernourishment and resource depletion in many ecosystems.

- Industrialism on the current fossil - fuel burning model is unsustainable in atmospheric terms.
- A certain amount of rapid urbanization is manageable, but not with the growth rates seen in cities such as Lagos, which grew by 10.2 million people in 1975 - 2000 at an average annual rate of 5.8% ( O' Meara, 1999).
- The 501 million cars in use worldwide are not only precipitating local crises of congestion and pollution, but adding to wider problems of environmental impact : the billion - strong fleet prediction for 2020 will contribute greatly to global pollution problems as well as to a growth in traffic across the world of some 60% in the period 2000-2020.
- Deforestation and carbon dioxide generation on a massive scale are eroding the earth's built-in adjustment mechanisms in many areas.
- Use of global fresh water and marine resources has grown tremendously : freshwater withdrawals have nearly doubled since 1960, such that humanity uses more than half the planet's accessible freshwater run-off—great rivers such as the Colorado and the Yellow River now fail to reach the sea much of the time as so much water is diverted for agriculture and industry; marine fish consumption doubled between 1960 and the early 1990s, such that some 60 per cent of the world's sea fish resources are over fished or at the limit of sustainable harvesting ( WWF, 1998).

Similarly, at the local level, the scale of new development is tipping many economies and ecosystems into crisis. A small number of tourists on a Greek or a Caribbean Island can be a boon to local life and even provide and economic basis for sustainable development. But when tourists outnumber local people by ten to one, and foreign travel companies package both local economy and local culture for sale, a threshold has been exceeded and negative effects begin to pile up for all concerned. In every case, the magnitude of change is too much. Critical, if unknown, thresholds have been exceeded and the situation is no longer amenable to beneficial local management. Usually, threshold for sustainability carrying capacities - are substantially exceeded even before we become aware of the nature of the problem.

The situation is made more complex and intractable because the trends and their consequences are highly interactive in a manner that is difficult to identify and measure, and sometimes even difficult to imagine. So, for example, urbanization is partly a result of, and partly a cause of, migration from countryside to city. The urbanization process itself generates economic activities which raise income levels, draw in resources from the countryside and even from far away Savannahs and rainforests leading to generation of enormous amounts of waste that end up as pollution of air, water and land. The increased income generates more consumption, more industry, more pollution, more automobility, urban sprawl, endemic traffic congestion, and so on. Urban sprawl results in the loss of prime farmland which, when combined with rural population growth, contributes to lowland forest loss due to agricultural expansion in the countryside well away from the city, thus completing a cycle of interaction. These processes are unfolding rapidly, but our responses to date fail to match the size of the problem.

Relationships between such trends and their consequences are not merely distinctions on a continuum of 'good' to 'bad' environmental effects : they are the very stuff of debate over the nature of sustainable development and the future of the planet. They go to the heart of our values and assumptions about how much of the earth's finite resources we are individually entitled to consume, and to our views on how much resource depletion is allowable in sustainable framework. They are also important because there has been little public debate about the meaning of sustainable development. It is perhaps most important to acknowledge that, given the political and economic constraints on environmental policy, progress towards sustainability will have to be incremental rather than revolutionary. This means that the best time to start doing something, and then to learn from what we are doing, is new.

## Natural Resources Assessment

Natural resources in a narrow sense are those uncaptured natural stores, which are useful to mankind in any way. They are most commonly grouped as waters, soil, forests, grasslands, wild animal life, and minerals. In so far as they constitute collectively one of the factors of production, the economist speaks of them as "land". Manipulated by human powers, which in turn are aided by tools and other operating devices, three factors, "land, labour (including enterprise), and capital, "swing into action, and what the economist calls "production" gets going. Forests become houses, railroad ties, fuel, paper, or pencils; grass and other forage become meat and leather; waters become available for drinking, cooking, cleansing, transporting and actually make up an important part of the volume of many useful articles and materials. Wild-animal life becomes food, fur coats, quarries, photographic targets, trophies, or museum specimens; minerals become household fuel, abrasives, fertilizers, paper clips, automobiles, and bulldozers (Singh, 2008). But the six groups mentioned above may well be arranged for convenience, and they might be lined up in this way :

1. Inexhaustible natural resources
   - The atmosphere
   - The water in its cycle
2. Replaceable and maintainable natural resources
   - Waters in place
   - Soils
     
     Land in its spatial sense
     
     For human activities
     
     For the scene and other amenities
   - Forests
   - Forage and other cover plants
   - Wild-animal life
   - Human powers
     
     Those of the body
     
     Those of the spirit
3. Irreplaceable natural resources
   - Minerals
     
     Metals
     
     Mineral fuels and lubricants
     
     Miscellaneous non-fuel, non-metallics
   - Land in natural condition
     
     Natural study areas
     
     Specimen wildernesses

Measuring sustainable Development (SD) involves valuation of environmental resources, damages due to exploitation and pollution and the costs of preventing and restoring the natural resource. Valuation of environmental resources and pollution effects needs for setting policy priorities and their implementation. In the absence of proper valuation, resources may be misallocated. Prices of goods indicate the scarcity of resources. But due to common property, nature of environment, there is free rider problem associated with their use. Those who do not pay can not be excluded from its use.

The imperatives of SD warrant maintenance of a constant natural capital stock (Sahu *et al.*, 2000). The valuation of natural resources is based on estimating its conservation value. The total economic value of natural resources includes present use and non-use value and future use and non-use value. Natural resource valuation involves techniques of contingent valuation; travel cost method and opportunity cost method. Use value can be estimated by using revealed preferences in conventional markets or in surrogate markets. Non-use value estimation individuals is difficult as this is derived on the basis of stated preferences of individuals in hypothetical markets.

Many environmentalists have criticized the term "sustainable development" as an oxymoron, claiming that economic policies based around concepts of growth and continued depletion of resources can not be sustainable since the term that implies resources remain constant. Resources such as petroleum are consumed much faster than they are created by natural processes, and are continually being depleted. It is argued that the term "Sustainable development" is a term invented by business to show capitalism as ecologically friendly, thereby placating people promoting environmentalist values (Singh, 2008).

However, technologies such as renewable energy, recycling and the provision of services can, if carried out appropriately, provide for growth in the economic sense, either without the use of limited resources, or by using a relatively small amount of resources with a small impact. In the latter case, even the use of small amounts of resources may be unsustainable if continued indefinitely.

Until 1980's the prime objective of the developing countries was to exploit the natural resources to the maximum extent so as to achieve rapid economic growth and increase in per capita consumption (PCI) of people. But with the UN Convention on Environment and Development, there is increased awareness among countries and people on the sustainability issues, hence the emergence of concern for environmental protection. The rational arrangement of natural resources is a necessary precondition for achieving economic growth and sustainable improvement in the living standards of the people. The natural capital is of fundamental importance in achieving changing production patterns with equity. The nation has entered a stage in which the existing resources will soon threaten the process of development. India is now on the threshold of a number of environmental problems, which if not solved in time, will adversely affect the productive capacity of the economy. India is an agrarian country producing agricultural goods, depends heavily on the management of natural resources, e.g., soil, water, vegetation and the climate. These elements are already beginning to suffer from considerable strain having an adverse effect on the quality and quantity of agricultural products. The relationship between growth equity and environmental sustainability is extremely complex. On the one hand, the transformation of natural resources into goods is essential for growth and quality of life, i.e., living standards. Thus, for example, the expansion of water supply is what makes it possible for growth of human settlements. On the other hand, however, there is the danger of development process affecting the quality of environment. Air and water pollution due to industrialization and concomitant social and economic change reduces the capacity of ecosystems to provide the community with vital goods and services. The theory of sustainable development deals with these issues and consequently, the changing priorities at length (Mohanty, 1999).

As duly recognized in Agenda 21, the urgent need for today is to utilize our natural resources in a sustainable manner with a focus on minimizing their depletion and pollution. Additionally, it was realized in Agenda 21, Chapter 13 that mountains are also important source of water, energy and biological diversity; accelerated soil erosion, landslides and rapid loss of habitat and genetic diversity often threaten these natural resources. Hence, the proper management of mountain resources and socio-economic development of the people deserve immediate action. The key elements of this theme are the utilization of natural resources at a sustainable level with a focus on minimizing depletion and the reduction in input of pollutants to these resources. The focal points are conservation of biodiversity, combating desertification and conservation of mountain ecosystems.

**Natural Resources Degradation**

The rising levels of production and consumption not only imply increasing resources flows, thereby leading to exhaustion of resources, but also increasing discharges of residuals, which if accumulated beyond the carrying capacity of the ecosystem. The resulting population impairs the ability of the natural environment to deliver useful flows of resources (Common, 1996). Environmental degradation of soil, water or air for example is understood as decline in the quality and/or quantity of the people's well being, in particular the livelihoods of the poor and their standard of living (Morvaridi, 1997). As per National Remote Sensing Agency (NRSA), 20.17 per cent of the geographical area of the country is categorized as degraded or wastelands (Kadekodi, 2004). Problems of water-logging and salinization have been reported in many case studies on evaluation of irrigation projects (Morvaridi, 1997; Chaudhary *et al.*, 1994). Though irrigation has played a major role in developing agriculture and improving income level of farmers, yet the rise in water table, excessive seepage from poorly lined canals and lack of drainage have contributed to water-logging and salinization. An evaluation of the socio-economic impact of environmental change that the irrigation project induced after its completion and utilization is carried out by Morvaridi (1997) in Sultanpur district of eastern UP on the Sarda Sahayak Irrigation Project. It was found that farmers have abandoned degraded lands as the costs of cultivating affected land proved to be higher than productive returns. The study showed that the cost of reclamation of degraded wasteland was Rs. 30,000 per ha for less severely degraded land to Rs. 50,000 per ha for more severely degraded land. Further, many farmers who could not make any investment on reclamation became landless by selling their land to large farmers.

Studies have shown that irrigation projects have negative socio-economic impacts. Numerous studies on development projects indicates that partial or only one - third of the total affected populations have been rehabilitated in India (Mahapatra, 1994; Fernandes and Asif, 1997). In another study in Orissa on development induced displacement reveals that the number of displaced persons during 1951 to 1990 was estimated to be around 30 million. While the '*patta*' (document of ownership) holders were compensated, other dependents were displaced without compensation or any other alternative to their livelihood (Fernandes, 2001).

Chaudhary *et al.* (1994) studied the environmental and socio-economic impact of irrigation projects in India in terms of regional and social inequalities, changing cropping pattern, loss of soil fertility leading to decrease in crop productivity, disruption in community life and health impacts. Disruption in community life has been observed in Srisilam dam in Andhra Pradesh, Kali project in Karnataka, Pong dam in Himachal Pradesh and Subarnarekha in Chhota Nagpur.

TERI's report on 'Green India - 2047' tells the state of problems of environmental damage and natural resources degradation in India. The report reveals that the pollution load on water resources is increasing due to burgeoning population, industrial growth and indiscriminate use of pesticides and other chemicals in agriculture. It is felt that approach to deal with issues related to water pollution has been ad hoc and sectoral so far (Pachauri and Sridharan, 1998).

Moreover, the quantity of solid waste generated is continuously on rise at an alarming rate and problems associated with it become more and more serious. Solid waste, if not managed effectively, adversely affects all the three components of the environments, namely air, water and soil. The main impacts of environmental degradation are as follows :

- ***Health Impacts***

Management and conservation of natural resources is important for the maintenance of health in addition to the need of food production and ecological considerations. Water pollution due to untreated sewage contributes to high coliform counts resulting in high infant morbidity and mortality. Gastroenteritis, a waterborne disease, is the first major cause of morbidity in India.

- ***Economic Impact***

It can be found in loss of availability of natural resources for the population, which is dependent on them for their livelihood. There could also be loss in crop productivity due to pollution or closure of industries to comply with air quality standards.

- ***Ecosystem Impacts***

There are in the form of contamination of ground water aquifers, land degradation due to water-logging, intrusion of seawater into ground aquifers, loss of water bodies, etc.

- ***Displacement Impacts***

Environmental degradation and non-availability of natural resources forces dependent livelihoods to migrate in search of alternate subsistence living.

- ***Aesthetic Impacts***

Loss of biodiversity - for example bird sanctuaries, lakes, etc., by visualizing the people get pleasure.

**Environment and Sustainability**

Environmental degradation is the damage to the biosphere as a whole due to human-induced activities. Environmental degradation occurs when nature's resources (such as water, air, earth, habitat and forest wealth) are being consumed faster than nature can replenish them, when pollution results in irreparable damage done to the environment or when human beings destroy or damage ecosystems in the process of development. Environmental degradation can take many forms including, but not limited to, desertification, deforestation, extinction and radioactivity. Some of the major causes of such degradation include : overpopulation, urban sprawl, industrial pollution, waste dumping, intensive farming, over fishing, industrialization, introduction of invasive species and a lack of environmental regulations.

This pollution has dramatic consequences for the health of the people. In the third world, pollution is generally the result of poverty, whereas in the industrialized countries, it is a by-product of prosperity and affluence. The goal of environmental sustainability is to minimize these, and other causes, to half and, ideally the process they lead to.

An unsustainable situation occurs when natural capital (the sum total of nature's resources) is used up faster than it can be replenished. Sustainability requires that human activity, at a minimum, only uses nature's resources at a rate at which they can be replenished naturally (Table 1).

Theoretically, the long-term final result of environmental degradation would result in local environments that are no longer able to sustain human populations to any degree. Such degradation on a global scale would, if not addressed, of course mean extinction for humanity.

In the short-term environmental degradation leads to declining standards of living, the extinctions of large numbers of species, health problems in the human population, conflicts, sometimes violent, between groups fighting for a dwindling resource, water scarcity and many other major problems.

**Table 1 : Consumption of resources and sustainability**

| *Consumption of Permeable Resources* | *State of Environment* | *Sustainability* |
|---|---|---|
| More than nature's ability to replenish | Environmental degradation | Not sustainable |
| Equal to nature's ability to replenish | Environmental equilibrium | Steady - State sustainability |
| Less than nature's ability to replenish | Environmental renewal | Sustainable development |

**Development: How Far Sustainable**

Previously, the sustainability of a human enterprise was measured solely in economic terms, starting with the United Nations Conference on the Human Environment held at Stockholm in 1972, environmental parameters were added to measure sustainability. The integration of ecological principles in economics took a concrete form in the shape of "Agenda - 21" adopted at the UN conference on Environment and Development held at Rio de Janeiro in 1992. Since then, considerations of environmental sustainability have been shaping the agenda for both technology development and policy formulation.

The growing rich - poor divide, jobless economic growth and the increasing feminization of poverty, all documented very well in UNDP's Human Development Reports of 1992, 1993, 1994 and 1995 have highlighted the fact that development which is not equitable will also not be sustainable in the long-term. Such inequitable development will lead to increased violence and promote social disintegration. Thus, we now know that tools for measuring sustainability will have to be enlarged in a manner that concurrent attention can be paid to economic viability, environmental sustainability and social equity. Equity will have to be measured both in gender and in economic terms. Such a paradigm shift in policy formulation in areas of concern both to science and society calls for a change in mindset among all professionals leading to viewing problems in a holistic manner. Such a change in mindset is essential for ending the prevailing mismatch between the capabilities of uni-directional minds and the needs of the multi-dimensional problems (Swaninathan, 1996).

Sustainable development is a collection of methods to create and sustain development, which seeks to relieve poverty, create equitable standards of living, satisfy the basic needs of all peoples, produce sustainable economic growth and set establish sustainable political practices. While taking the steps necessary to avoid irreversible damages to natural capital in the long-term, in turn, for short-term benefits by reconciling development projects with the regenerative capacity of the natural environment. The field of sustainable development can be conceptually broken into four constituent parts, viz., environmental, economic, social and political sustainability.

**Need for Sustainable Development**

Improving human well being over time is a broader goal the increasing economic growth that focuses primarily on material comforts. This has some important implications. Since social and environmental aspects also affects human well being directly, a strict policy of 'grow now, clean up later' involves costs for today's generation, costs that often fall disproportionately on 'today's poor.' Durable economic growth is required for a serious attempt at poverty reduction. This means paying enough attention to social and environmental concerns so that durable growth is not affected. There are limits to sustainability among assets, perhaps even more from the perspective of people's well being. When environmental or natural assets are fairly by the latter can be expected to lead to higher returns. Development strategy till date has often relied on drawing down environmental resources and replacing them with human made assets. Most developing countries' growth strategy continues to focus largely on the accumulation of physical assets (Manikandan and Rao, 2005).

The way an economy grows pace and pattern of growth - can matter for the well being of both the current generations children and grand children. Developing countries do not have to follow the path of development traversed in the industrial countries.

**Indicators for Sustainability**

Theoretically, one may say that sustainability is a process of change in which exploitation of resources, direction of investments, orientation of technological changes etc., are made consistent with the future as well as the present. We may say that economic development in a specified area i.e., a region or a national or throughout the globe is sustainable if the total stock of resources human capital, physical reproducible capital, environmental resources, exhaustible resources, does not decrease overtime. The most objective and potentially measurable criterion for sustainable development is the preservation of the productively and full functioning of the natural resource base (Raymond, 1992).

One may define 'sustainable development' as the preservation of the production possibilities of an economy to provide the same goods and services including services obtained from the state of nature (Parikh, 1989). The difficulty with this approach of defining sustainability lies with its static nature. The consumption pattern in future might change with improvement in the standard of living, causing demand for a variety of goods and services. The economic system has to meet with that, so we modify the above definition by saying that expanding "Production Possibility Front" implies sustainability of the economy. But, this is a partial approach. This is not explaining the environmental dimension explicitly.

When we consider the total development process, the problem of environmental pollution and conflict between industrial development and environmental degradation become very serious. At this stage merely the technical side of production can not be blamed for environmental pollution. The consumption pattern of society, the living styles of people, the social and political values of the masses, attitudes, customs etc., all reinforce the environmental degradation. For sustainable development, who have to take all such factors into account along with the technology (Barthwal and Shukla, 1994). The United Nations General Assembly, through its Agenda 21 (UNCED, 1992) has provided a comprehensive picture of the interlinks related to environment and sustainable development, i.e., the priority of the goals of the economy for sustainable development and the means to achieve them. Accordingly, one has to link national and international policies for revitalizing economic growth with sustainability. Combating poverty, improvement in demographic structure, changes in consumption patterns, health, human settlements, pollution control, energy management, treatment of industrial wastes, control of hazardous materials, and after all the input sustainability are the vital requirements for overall sustainable development of nations.

From the ongoing discussion, it is clear that achieving sustainable development is a complex and difficult task. One has to keep in mind that the ultimate goal of sustainable development is to maintain a rising trend in the welfare of the people, of the present and future generations. Sustainability links present to the future. One way is to develop dynamic mathematical models linking all dimensions of environment, economics and socio-cultural changes. The sustainability conditions can be derived in objective terms from such models. However, this will be complex exercise, full of difficulties. Therefore, from operational point of view, we need simple indicators to assess environmental and economic sustainability of the development process. Numerous economists/environmentalists have suggested such indicators (El Serafy, 1989, Goodland, *et al.,* 1991, World Bank, 1991, Holmberg, 1991, Mathews and Tunstall, 1991, Daly and Cobb, 1989, Dalai Clayton, 1992, Clarke, 1991). In fact, the whole literature on Environmental Impact Analysis provides excellent guidelines and indicators for sustainability of economic and ecological system (Rau and Wooten, 1985, Ryding, 1992).

At present, sustained increase in GDP or per capita income is the most used measure of development, but, as we know, it fails to allow for capital maintenance of natural assets and takes very limited account of the environmental contribution to economic activity. So in the context of sustainable development, it is highly inadequate and should be replaced by a set of indicators reflecting changes in the environment and economic activities. The environmental indicators for measuring changes in the state of the environment should be simple and practical. They might be expressed in non-monetized or physical terms. These indicators are to be compared with their sustainable limits. The other set of indicators should reflect the progress towards the broader goals of sustainable development in the national context. Both economic and social changes should be considered by such indicators. Some important indicators of sustainable development are as follows :

**1. GDP Growth Rate**

This reflects the success or failure of the economy as a whole. High growth rate of GDP is essential for providing employment, price stability and material welfare of the people. It helps in elimination of poverty, which in turn reduces environmental pollution caused by poverty.

**2. Population Stability**

Human population of a country is a source of all economic activities. Unfortunately, it is also a major source of environmental pollution. So there is an urgent need to control population growth. No net increase in population over some planning period is highly desirable. In fact, in several parts of the world it has to be negative.

**3. Human Resource Development Index**

The Human Development Index (HDI) is a comparative measure of life expectancy, literacy, education and standard of living for countries worldwide. It is a standard means of measuring well - being, especially child welfare. It is used to determine and indicate whether a country is a developed, developing, or underdeveloped country and also to measure the impact of economic policies on quality of life. The index was developed in 1990 by Indian Nobel prize winner Amartya Sen, Pakistani economist Mahbub ul Haq, with the help of Gustav Rannis of Yale University and Lord Meghnah Desai of the London School of Economics and has been used since then by the United Nations Development Programme in its Annual Human Development Report. An index showing high magnitude, i.e., more than 0.8 is considered to be quite high and less than 0.5 as poor for assessing HDI. The UNDP occasionally publishes the magnitudes of this index for different countries.

The HDI measures the average achievements in a country in three basic dimensions of human development : (a) a long and healthy life as measured by life expectancy at birth; (b) knowledge, as measured by the adult literacy rate (with two-thirds weight) and the combined primary, secondary and tertiary gross enrolment ratio (with one - third weight); and (c) a decent standard of living, as measured by the log of Gross Domestic Product (GDP) per capita at purchasing power parity (PPP) in US$.

**4. 'Clean Air' Index**

The ambient air should be 'clean' for survival of human beings and other species. But, the level of air pollution is increasing day by day with the pace of development. Several gases like, sulfur dioxide, carbon monoxide, carbon dioxide, oxides of nitrogen ( $No_x$ ), suspended particulates, hydrocarbons, etc. pollute air quality to a great extent. Some of them are responsible for the green house effect while some are responsible for acid rains, which are quite harmful for life. Toxic substances in the air, on the other hand, create several health problems. The adverse effects of air pollution should be minimized for the sustainability of development.

**5. Energy**

Energy is vital input for development. It is, therefore, quite important to see the impacts of energy supply and demands for sustainable development. In general, the energy intensity of aggregate output measured by energy / GNP ratio has to be lower for greater sustainability.

Energy issues are an important part of sustainable future. One of the major interests of the Earth Institute is the study of energy and the ways that it is connected to climate change, poverty alleviation and sustainable development. Alternate energy sources may be the viable option in order to develop future energy systems that help define the future of energy for our planet.

**6. Water**

The supply of adequate quantity of water for domestic consumption and various industrial and agricultural uses is to be ensured for sustainability.

**7. Land Degradation**

Loss of top fertile soil leading to depletion in soil productivity is to be reduced to minimum level for sustainable development.

**8. Forest cover**

India is one of the 12 mega diversity countries having a vast variety of flora and fauna, commands 7% of world's biodiversity and supports 16 major forest types, varying from the alpine pastures in the Himalayas to temperate, sub-tropical forests, and mangroves in the coastal areas. According to the state of Forest Report, published by the Forest Survey of India (FSI) in 1997, India has a recorded forest area of 76.5 million hectare or 23.3 per cent of the total geographical area of the country. Out of this area, a total of 31.85 million ha forest in the country are degraded or open.

Destruction of forests means loss of water, climate, soils and even wildlife. So maintenance of appropriate forest coverage ratio for sustainability of economic development is must for a secured future.

Several other indicators like housing conditions, common property resources, crime rate, cultural stability, social tensions, etc. may be added to this list. In addition, prices of resources used will be most reliable indicators for the sustainability.

## Natural Resources and Economic Development Nexus

The issue of use and misuse of natural resources by human beings is inextricably linked to the economic development. The lives and livelihoods of citizens in rural habitats and urban agglomerations are often dependent on their natural surroundings. It is chiefly because occupations and/or economic activities are carried by human beings on the earth. Therefore, the rate and path of economic development are determined by the existence and use of the natural resources.

Every nation - rich or poor, agricultural or industrial - has to take care of the four major resources of the Earth, viz., land, water, air and forests.

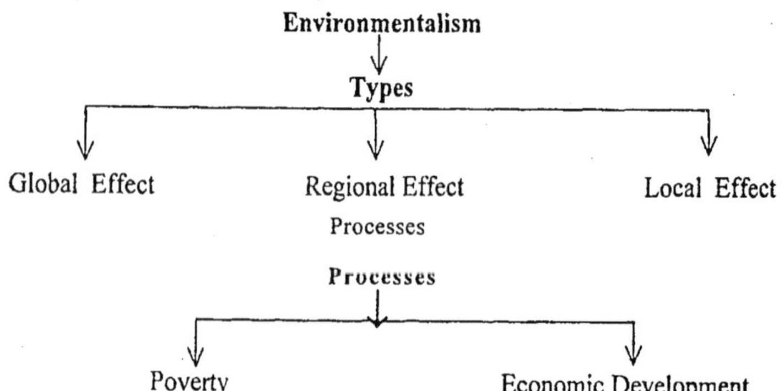

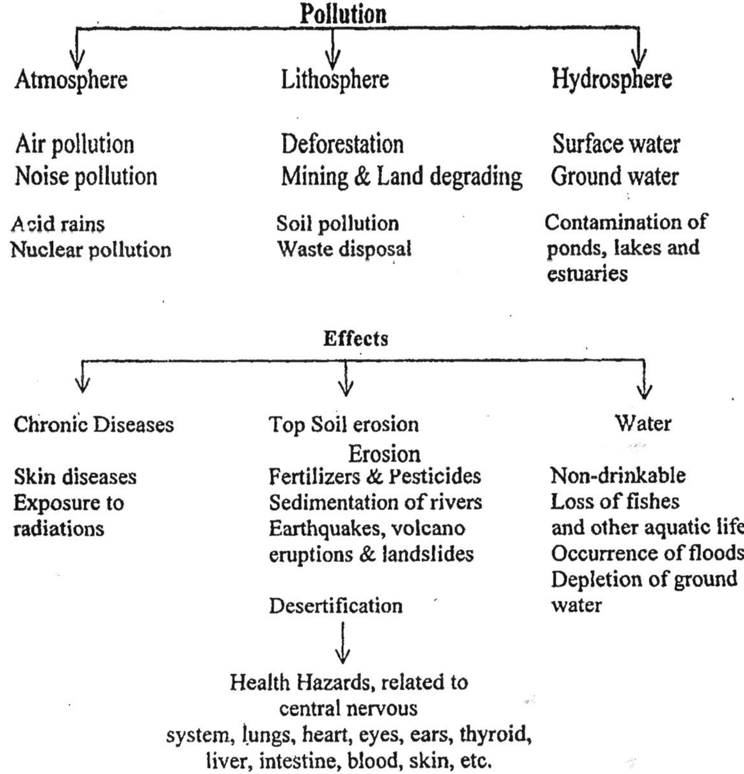


**Source:** Moolya, 1998

The use of natural resources has to be guided by the principle of optimum utilization, in the short as well as the long-run. This is not only an international agreement but also an individual commitment. The rich North consumes 83 per cent of the world output and the affluence causes negative externalities in abundance. For instance, the affluent US alone releases about 28 per cent of chlorofluorocarbon (CFC) into the environment. On the other hand, the poor South is so laden with debt that it is forced to over exploit its resources and sell it to rich North. The glaring inequality in wealth consumption today is therefore, quite a challenge : how can the rich nations maintain their economic growth alongside environmental management, and how can the poor nations attain faster growth without degrading the environment. The ecological imbalance has come from both sides. Both have to address the challenge. With new incentives, impulses and implications of liberal and global economic growth, the issues tend to become more complicated. This is true for India too.

**Conservation and Management of Natural Resources**

The state of prime natural resources of the country like India is described below:

*1. Land Use Planning*

The role of natural resources in economic development need not to be emphasized. The availability or non-availability of favorable natural resources can facilitate or retard the process of economic development. While pointing out the importance of natural resources in economic development. W. A. Lewis writes, 'Natural resources determine the course of development and constitute the challenge which may not be accepted by human mind'. Fisher writes, 'Usually have to beg in with and concentrate on the development of locally available natural resources as an initial condition for lifting local levels of living and purchasing power, for obtaining foreign exchange with which to purchase capital equipment and for setting in motion the development process. Natural resources that indicate potential wealth of a country mainly include land water, fisheries, mineral, forests, climate, rainfall, topography and marine resources. The poor management of natural resources constrains the development efforts in many areas and economic activity will pose a serious challenge for environmental management.

In the third world countries today including India the major environmental problems are associated with the misuse of natural resources like soil, forests, and water resources, which are the consequence of the contemporary development paradigm. Our existing paradigm of development not only contributes to the depletion and degradation of the natural resources but also accentuates the problems of inequality, unemployment and poverty. Hence, the concept of 'sustainable development' for the developing countries like India need to be governed by the principles of conserving and preserving the existing natural resource base and mitigating the problem of inequality, unemployment and poverty. The major challenge faced by India today is to evolve a development process that will lead to greater equity, growth and sustainability. To achieve the goal of sustainable development, full participation is must for the management of natural resources. If all the nations choose to participate to the international environmental agreements, the problem of mismanagement of natural resources may be tackled leading to ensured sustainable development.

Land is the most precious asset and non-renewable endowment. Conservation of land resources can promote sound land compatible to the land capabilities. In India, nearly 70 per cent of the population is dependent on agriculture for the livelihood. It contributes 25 per cent of the Gross Domestic Product (GDP). The land under agricultural use has increased to the extent of 9 per cent during the period 1958-59 to 1990-2000. The area under non-agricultural use has also increased to a great extent. There is a significant decline in the area under barren and uncultivable land, miscellaneous tree and groves, cultivable waste, permanent pastures and grazing and other fallows. The increasing population of human and livestock, poverty and rapid economic development are exerting heavy pressure on the land. This has ultimately led to very significant land degradation. From viewpoint of ecological development, it is important to understand the extent and magnitude of wastelands also (Kadekodi, 2004). According to Sehgal and Abrol (1994) 57 per cent of the total geographical area of India has been degraded. According to National Remote Sensing Agency (NRSA), the total wastelands in India during 1999 stood at 64 million hectares as against 44 million hectares in 1998-99. This amounts to a sharp rise in wastelands by almost 44 per cent over 10 years (Kadekodi, 2004).

To conserve the land resource, proper planning is essential. Sustainable land use system can be seen as a farm system that interact with biophysical and socio-economic environment in such a way that they are economically viable in short run and ecologically sustainable in the long-run. To achieve sustainability, proper planning is imperative. For land use planning the resource of the region should be appropriately and systematically documented, mapped and potentials and constraints should be fully assessed. In other words, it can be called as resource appraisal and can be defined as a systematic characterization and mapping of biophysical resources, viz., soil, climate, land reforms, vegetation, surface and ground water, transport network and village settlement, land use, etc.

Using traditional and remote sensing techniques and correlating and integrating with socio-economic data, planning for land use zones may be formulated. Land use planning has to incorporate sustainability in both its policies as in its implementations. The planning process must be tailored as a dialogue between the overall environment and landscape intentions and the actual local ambitions for productions and social developments. Thus, sustainable land use planning can be the instruments to set land use polices and improvements of the spatial area for an optimal use and protection of the natural resources on the long-term while meeting the needs and aspirations of the present generations.

*2. Forest*

Forests are a very valuable remarkable resource, providing a vital life support system. However, in recent years the area under forest cover in our country is declining at a faster pace due to many reasons. Indiscriminate felling of trees due to demographic and commercial pressures, debasement of permanent pastures due to overgrazing and accelerated rate of deforestation are the major reasons behind the state of forests. These, in fact, are not only a bane on the ecological balance but an affliction on the sustainable development of our economy also.

India is having 675538 km$^2$ of forest cover and it constitutes 20.55 per cent of its geographical area. The dense forest is 416809 km$^2$ (12.68%) and open forest 258729 km$^2$ (7.87%). It is surprising that the forest cover has increased to an extent of 38245 km$^2$ (6%) during the year 1999 and 2001 (GOI, 2004). According to the State of Forest Report 1997, India has lost 0.548 million hectares of forest cover during 1993-95. Again during 1995 and 1997, the forest cover went down from 63.89 million hectare to 63.34 million hectare. Dense forests decreased by 1.78 million hectare during 1991-95. On December 12, 1996, the Supreme Court banned illegal felling of trees in all natural forests (except those under state working plan). According to the State of Forest Report (1995), *Jhum* cultivation and non-regeneration of degraded forests are the main causes of depletion of forests in the country. But, the 1997 report does not analyze the causes of forest destruction (Agarwal *et al.*, 1999).

Depletion of India's forest wealth has brought into focus an urgent need for forest management. Sustaining the forest resource base against the growing human and livestock population pressure, urbanization and industrialization have become the need of the hour. Apart from these pressures, the dependence of forest users groups is also a crucial factor in condition of India's forest.

The sustainable development of forest can only occur through community participation leading to the concept of Joint Forest Management (JFM). The JFM can be a negotiation process between the local communities and the Department of Forest to protect the forest jointly. With government's effort, JFM provides for community protection of forest by assigning to the villagers a forest guard and a regular watchman in addition to the special watch being kept by each member by rotation to protect the regenerating areas from theft, illicit cutting, fire and encroachment. However, the JFM can only be successful only when local people's effective participation is ensured through education, awareness building, motivation and persuasion.

For sustainable use of forests, one has to regenerate the forest on a continuous basis and this can be done by regular and continuous flow of forest products and other specific services, which may be essential to sustain the livelihood of the forest dwellers. Also the long-term viability of alternative uses that replaces the original ecosystem should be there.

### 3. Water Resources

India's landmass and water resources account for about 2.4 per cent and 2.5 per cent, respectively of the world while 16 per cent of the world population in India is dependent on these scarce resources. In India, the average run-off of water in the river system is 1869 cubic kilometers whereas its utilizable portion is about 690 cubic kilometers. The potential is estimated at 432 cubic kilometres. The annual per capita availability of water fell from 6008 cubic metres in 1947 to 2464 cubic metres in 1990 to 2266 cubic metres in 1997 (Pachauri and Sridharan, 1998) and 1998 cubic metres in 2001 with great variation in water availability in different river basins (GOI, 2004). At the end of 9$^{th}$ plan, India's irrigation potential has been increased to about 93.98 million ha from 22.6 million ha in 1951 (GOI, 2004), an increase of 31.6 per cent during a period of about 50 years.

All the rivers in India are polluted with toxic substances The main causes of water pollution are urbanization, industrialization, over utilization of irrigation water leading to water-logging and salinization, agricultural run-off with accumulation of pesticides and discharge of domestic untreated wastes, disposal of corpse and other waste materials into the rivers. Most of the cities in India, depend on rivers for their drinking water requirements (Agarwal *et al.*, 1999). But, the growing pollution of our river constitutes the biggest threat to public health. During 1990's more than one million children died due to diarrhoea and other gastrointestinal disorders and a total of 50 million children have died so far since independence. Now, the adults are likely to suffer and die of horrendous diseases because of the chemical pollution/loading (Agarwal *et al.*, 1999).

### 4. Minerals

Minerals are non-renewable and their further growth and regeneration is not possible. Currently, 2000 known deposits with as many as 87 minerals (including 4 fuels, 11 metallic, 50 non-metallic and 22 minor minerals) are exploited in the country (MOEF, 2002). Thus, our mineral consumption is increasing day by day and soon these resources will be exhausted. Mining activities necessarily lead to land degradation. Data related to area actually affected by mining and quarrying are not available. The mining lease area is about 0.8 million hectares in India (MOEF, 2001). Mining activities necessarily lead to land degradation, pollution of air and water, noise pollution, loss of vegetation and land degradation.

Alongwith the conservation practice, we have to think also to minimize the adverse effects of mining and their processing. A reduction of the adverse effects on the environment and its compounds should be considered at the initial stages of exploration and development for mineral extraction. Systematic and planned mining operations, proper site selection for waste disposal and tailing, introduction of no waste or minimum waste technology, use of sophisticated technology for exploitation would reduce the adverse effects to large extent. Afforestation in wasteland as backfilled pits will help in rehabilitation and will enable sustainable development of mineral resources in future.

Sustainable development aims at meeting the needs of the present without compromising with the ability of future generations to meet their own needs. Thus, sustainable' development requires that the rate of depletion of new renewable resources should foreclose as few future options as possible. Sustainable development requires that the adverse impacts on the quality of air, water and other natural elements are minimized so as to sustain the ecosystem's overall integrity. In order to achieve a harmonious equilibrium between the imperatives of mineral development and those of preservation of the environment, India's National Mineral Policy, 1993 has spelt out protection of environment as one of its major objectives (Suri and Rawat, 1996).

*5. **Biodiversity***

In India, many plants and animal species are threatened by a destruction of their habitat and over-exploitation of resources. A large number of species are either endangered or on the verge of extinction, both of which can by attributed to a lack of policy and institutional mechanisms (MOEF, 2001). Biodiversity supports human survival through health, food and industry and it has social, ethical, cultural and economic values (Joshi and Joshi, 2004).

India ratified the International Convention of Biological Diversity on February 18, 1994 and became party to the convention in May, 1994. The Convention of Biological Diversity (CBD) is an international legal instrument for the conservation and sustainable use of biological diversity taking into account "the need to share cost and benefit between developed and developing countries and the ways and means to support innovation by local people". It was resolved to evolve an international regime on access to genetic resources and benefit sharing with the aim of adopting an instrument instruments to effectively implement the provision of CBD (MOEF, 2005).

The Biological Diversity Act, 2002 implemented by the Indian Government incorporates provision laid down for the conservation and protection of our rich biological diversity.

**National Policies and Legal Framework : Approach Towards Conservation and Protection of Natural Resources**

In India, the existing policy framework for environmental protection rely largely on command and control approach while the developed countries favor market-based instruments. A series of environmental legislations were introduced after UN Conference on Human Environment held at Stockholm in 1972. Until then the awareness and concern on environmental issues were negligent among the public and policy makers. Major initiatives taken by Government of India during the economic reforms period in the background of the constitutional framework are as follows (Box I).

**Box I: Natural Resources and Environment as enshrined in the Indian Constitution**

- Article 39 (B) provides that the ownership and the control of material resources of the community are so distributed as best to subserve the common good.
- Article 48 (A) says that the State shall endeavour to protect and improve the environment and to safeguard the forests and wildlife of the country.
- Article 51 (A) specifies that it shall be duty of every citizen of India "to protect and improve the natural environment including forests, lakes, rivers and wildlife, and to have compassion for living creatures."
- 42nd Amendment Act, section 10 says, "The state shall endeavour to protect and improve the environment and to safeguard the forests and wildlife of the country."
- The 73rd and 74th amendments to the Constitution recognize the role of local bodies in environmental protection and include protection of the environment and promotion of ecological aspects to be taken up by local bodies.
- The Eleventh Schedule (Article 234 G) entrusts in addition to others (total 29), the following functions related to natural resource management to *Panchayats* :

(i) Agriculture, including extension,

(ii) Land development, implementation of land reforms, land consolidation and soil conservation,

(iii) Minor irrigation, water management and watershed development,

(iv) Animal husbandry, dairying, and poultry,

(v) Fisheries,

(vi) Minor forest produce,

(vii) Drinking water,

(viii) Fuel and fodder,

(ix) Non-conventional energy sources, and

(x) Maintenance of community assets.

The Twelfth Schedule (Article 243 G) lists in addition to others (total 18), the following functions related to natural resource management to municipal bodies (Ministry of Law and Civil Affairs, 1996).

(i) Water supply for domestic, industrial and commercial purposes,

(ii) Solid waste management,

(iii) Urban forestry, protection of the environment and promotion of ecological aspects, and

(iv) Cattle pounds, prevention of cruelty to animals

***Source*** : Ministry of Law and Civil Affairs, 1996

**National Policies/Programmes**

National Forest Policy (NFP), 1988 lays down "village and community lands including those of foreshores and environs of tanks not required for other productive uses should be taken up for development of tree crops and fodder resources".

The Policy states that revenue gathered through such programmes should belong to *Panchayats* where lands are vested in them. In all other cases, local community should share such revenue in order to provide an incentive to them (Singh and Lal, 2001). The National Policy came into enforcement in April, 1992.

The NFP 1988 recognized the needs of forest - based livelihoods. It was realized that natural resource management is not single - handed and involving those who depend on them in conservation would benefit both the livelihoods as well as development of resources. The concept of decentralized management of forest resources got impetus.

The Ministry of Environment & Forest has come out with the National Forestry Action Programme (NFAP) a framework providing guidelines for sustainable development of forests. NFAP thrusts on following issues :

- Protection of existing resources, i.e., protection and conservation of soil, water and biodiversity,
- Improving forest productivity through rehabilitation of degraded forests,
- Reduction of demand on forests through substitution and improvements through introduction of technology,
- Strengthening Policy and Institutional Framework through capacity building, forest policy, legislation, public forest administration, research, planning and budgeting, and
- Expansion of forest area through tree plantations and people's participation in development and conservation.

The policy statement for abatement of pollution issued by the Ministry of Environment & Forests expresses concern over deterioration of environment, and advocates a mix of instruments for environmental protection. Polluter pay principle, public role in decision making and market - based instruments in fixing the cost natural resources so that realize their value are the main features of this policy.

By promulgating Rule 14, under Environment (Protection) Act, 1986 the Government has made it compulsory practically for every industry to conduct an environmental audit and submit environmental statement to the concerned State Pollution Control Boards (SPCBs) every year (Bengeri, 1995).

The Report of the Technical Committee on Drought Prone Area Programme and Desert Development Programme (1994) says that "Since the objectives of watershed development is ecological improvement and conservation of natural resources as well as socio-economic development of local population, watershed management efforts must incorporate soil and water conservation and land use planning into a broader framework that takes into consideration not only physical interrelationships but economic, social and institutional factors as well and appropriate arrangements should be made for distribution of benefits to watershed community from the government lands and watershed development". (Singh and Lal, 2001).

National Water Policy, 2002 states that "In view of the vital importance of water for human and animal life, for maintaining ecological balance and for economic and developmental activities of all kinds, and considering its increasing scarcity, the planning and management of this resource and its optimal, economical and equitable use has become a matter of utmost urgency. Concerns of community needs to be taken into account for water resources development and management".

India's perspectives on sustainable development have been presented in the Study Report "Empowering People for Sustainable Development" (EPSD) 2002. The Report was presented in World Summit for Sustainable Development (WSSD) in 2002. India's commitment to the process of sustainable development has been through strategies for overall development stated in five-year plans. But, EPSD has evolved a holistic approach for sustainable development casing all the sectors following a multi-dimensional approach. Conservation of natural resources is taken up as a priority issue as it sustains around two-third of the country's livelihoods and contributes to environmental sustainability.

**World Summit on Sustainable Development (WSSD), Johannesburg (2002)**

This meeting reviewed progress on sustainable development made since the landmark 1992 Earth Summit held in Rio de Janeiro, Brazil. It was held in Johannesburg, South Africa from September 2-11, 2002 and attended by more than 110 heads of State, as well as over 50,000 delegates and members of non-governmental organizations (NGOs).

Preparative committee on WSSD emphasized largely on 'new development models'. "New ways and means would have to be found to move towards sustainable development", stated Emir Salim of Indonesia who was chairing the preparatory committee for WSSD. "A new ethics, based on recognition that the current model of development is outlined, is required to meet the challenges of the new century, "stated Crappie Tickell, a former ambassador of UK to the UN, at a WSSD preparatory roundtable of experts held in Vail, Colorado, in June 2001. Key issues at WSSD were:

***(a) Poverty***

An Asian and the Pacific Ministerial Conference held in Kitakyshu, Japan, from September 4-5, 2000, identified, "the poverty situation in the region" as a major obstacle to sustainable development and recommended that integrated and participatory action plans in adequate details be drawn up and implemented to alleviate poverty.

***(b) Globalization***

The main question is how to make globalization work for sustainable development. Another preparatory round table for African held in Cairo, Egypt (June 25-27, 2001) stated that "it is hoped that WSSD will seriously address the issues of globalization and contribute to the steps of sustainable development."

***(c) Financing for development***

There have been few recommendations on "innovative financing". The roundtable for East Asia and the Pacific called for engaging in policies based on the "polluter pays principle". The Central and South Asian roundtable, on the other hand, recommended that WSSD call for reduction of military spending and redirect such expenditure to social and environment programmes.

Surprisingly, there seem to be few takers from alternative financing mechanisms such as the Tobin tax. The Noble Laureate Economist James Tobin originally suggested the Tobin tax, which has attracted attention in the international arena in recent years. It was basically a tax on very short, two-way international currency transactions.

***(d) Fresh water***

Privatization of water and the relationship between the private and public sectors in water management was a key focus of the WSSD discussions. Northern countries seem to be particularly pushing the agenda of privatizing water, identifying it as an economic good. The logistics of this proposal such as identifying the role of stakeholders still remain hazy.

***(e) Governance***

There have been calls from certain sectors for creation of a world Environment Organization (WED), along the lines of the World Trade Organization (WTO) that oversees the global multi-lateral trading framework, to spearhead international environmental governance.

Kofi Annan, UN Secretary General, has stated that WSSD will be a golden opportunity for world leaders to show that they take the idea of stewardship of environmental governance seriously. Its success depends on the willingness of national governments to go beyond more rhetoric. That was the biggest hurdle and will also be of WSSD of the Earth Summit (Rana, 2005).

## SUSTAINABLE AGRICULTURE

"Sustainable development alongwith sustainable agriculture is necessary for the execution of an integrated approach to increase the food production and enhancement of food security and food safety in an environmentally sustainable way". In fact both the concepts are interdependent and interrelated. To create the conditions for sustainable agriculture major adjustments are needed in agricultural, environmental and macro-economic policy at both national and regional levels.

"Sustainable Agriculture" according to Lockeretz (1988) is a loosely defined term that encompasses a range of strategies for addressing many of the problems that afflict agriculture. These problems include loss of soil productivity from excessive erosion/intensive cultivation and associated plant nutrient losses/depletion : surface and ground water pollution from pesticides, fertilizers and sediments; impending shortages of non-renewable resources; and low farm income due to low commodity prices and high production cost.

The agricultural chemicals by influencing soil flora and fauna severely impair the nitrogen fixation process as well as degradation of plant wastes because these chemicals are toxic to soil micro-organisms including *Rhizobium*. Thus, unless toxicity/residue levels are determined, these pesticides should not be used or efficient microbial strains be developed which are capable of degrading these compounds so that proper health can be maintained (Singh and Verma, 2000).

The priority must be on maintaining and improving the capacity of the higher potential agricultural lands to support an expanding population. The main tools of sustainable agriculture are policy and agrarian reform, soil conservation and improved management of inputs.

Sustainable agriculture also emphasize on the need for linkages between economic and ecological indicators of change in land use. Possible approaches towards Sustainable Agriculture are - (a) Crop diversification, (b) Genetic diversity, (c) Integrated Nutrient Management (INM), (d) Integrated Pest Management (IPM), (e) Soil suppressiveness, (f)

Efficient Water Management, (g) Post-harvest Management, (h) Energy Management, and (i) Extension of Technologies and Managing Information Input (Dayanandan, 2005).

Sustainable agricultural strategy is to find out alternative route for achieving the goal without harming the long-term productive potential of soil. According to the American Society of Agronomy (1989) a sustainable agriculture is one that, over the long-term, (i) enhances environmental quality and the resource base on which the agriculture depends, (ii) provides for human food and fiber needs, (iii) is economically viable, and (iv) enhances the quality of life for farmers and society as a whole.

The ultimate goal or the ends of sustainable agriculture include : (i) developing farming systems that are productive and profitable, (ii) conserving the natural resource base, (iii) protecting the environment, and (iv) enhancing health and safety.

## NATURAL RESOURCE MANAGEMENT : PARTICIPATORY APPROACH

Natural Resources can not be managed and preserved properly without involving the rural communities, which derive their sustenance from natural resources. Participatory natural resources management was a part of community life in the past. The community maintained and preserved 'sacred groves', village tanks/wells and pasture lands, though these are traditionally owned by the State. But, later the increased government's intervention added by its failure to preserve the natural resources and their exploitation for the sake of development projects and unregulated use as free good by the commons led to environmental degradation. The worsened situation has compelled people, particularly the villagers and tribes to come forward for the conservation of natural resources for sustainable development because ultimately it is they who suffer due to loss of natural resources (Mishra and Bajpai, 2001).

Murthy (1994) observed that in developing countries, the people's capabilities to harness the preserved common and the fairness in the appropriation of benefits from them are the limits on the voluntary collective action. He argues that collective action is possible if an outside agency plays the role of a catalyst in mitigating these limits. However, population growth turns common and state poverty resources into open access resources. A participatory approach to development is essential as it ensures the proper identification, utilization and preservation of resources and also the reduction of the alienation of the people from the development process (Sethi, 1996). Further, it also helps to reduce the power of the bureaucracy and makes development planning people inclusive.

The documentation of the recent initiatives in participatory management of natural resources indicates that there is increasing pressure for conserving natural resources for the common well-being. Examples of such initiatives particularly with reference to water and forest resources as documented in research studies in India are : *Van Panchayats*, Sukhomajari Experiment, Joint Forest Management (JFM), Ralegaon Siddi experiment, *Phad* system and watershed development.

### 1. Forest Resource Management by Community

Community based conservation of natural resources is necessary from the view of both conservation and social justice, and local people, because of their day-to-day interaction and dependence on natural resources are often at the forefront of protest against the degradation caused by communal interest (Kothari *et al.*, 1998).

#### *(a) Joint Forest Management*

The Joint Forest Management (JFM) was initiated in 1990. This programme promotes sharing of products, responsibility, control and decision-making authority over forest land between the forest department and local users. JFM indicates fundamental shift from centralized management to decentralized management of forest, revenue generation to resource development orientation, single benefit to multiple benefits, monoculture to multiple cropping, and above all unilateral decision to participatory decision making (Poffenberger *et al.*, 1996; Dhangare, 2000). Twenty seven states have implemented the JFM programme and over 18 per cent of country's forestland is under this programme. In absolute terms, the area under JFM is more than 14 million hectares involving around 62,890 JFM groups (Sehgal, 2001). There are evidences from several studies on JFM that the programme has resulted in the improvement in the condition of the forests. There are reports about reduction in area under encroachment and fall in the rate of fresh encroachment, and increase in the income of participating communities at several places. However, Sunder *et al.* (2001) opines that JFM is too diverse to allow generalized conclusions about its success or reliability.

#### *(b) Van Panchayats/Forest Panchayats*

Another form of local community management of forest resources, which have legal basis, and clearly defined duties and powers. They were evolved in Uttar Pradesh in 1920s and 1930s out of bitter conflicts between local communities and the forest department. *Van Panchayats* (VPs) were created to give elected representatives of local people a well defined role in the management of civil (unreserved) forests under the overall supervision of district magistrates. The VPs have responsibilities for maintenance and preservation of forests, distribution of produce on the basis of local demand and Forests Department's approval, and execution of forest department plan. It is reported that VPs are managing about 15 per cent of the forest area in the UP hills. Despite their remarkable growth and good record in the village forest management, the VPs have been facing troubles in their functioning by unnecessary administrative controls and not able to realize their full potential for the development of forestry in the hills. The success of VPs in the hills has been attributed to their ability to meet people's basic needs from forests on an equitable basis, the homogeneity of caste composition amongst hill people, and fairly egalitarian distribution of land among the families in the village.

## 2. Participatory Management of Water Resources

The National Water Policy, 2002 emphasizes the role of Water User Associations (WUAs) and *Panchayati Raj* Institutions (PRIs) in the operation, maintenance and management of water resources infrastructures and facilities. It gives direct impetus to community participation in all facets of water management. The policy recognizes the need to "harness every drop of rainwater" as a national priority with special emphasis on decentralized harnessing of water resources with active community participation.

Sustainable water resource management rests on adopting a comprehensive water resource management with due considerations to environmental management, viewing water as an economic good and introduction of technological developments in water resource management (Raju, 2002).

Water management initiatives can be successful only if all categories of farmers accept responsibilities and ownership of the interventions. WUAs which perform agriculture-related functions in addition to water management and undertake some group activity on their own before getting any assistance from government are likely to be more sustained, active and effective (Hooja, 2002). The approach of the international funding agencies and the introduction of 73$^{rd}$ and 74$^{th}$ amendments to the constitution has brought in changes in government policies in many sectors. Decentralization, devolution of powers, participatory and bottom up approach are the new words found in recent government reports, policy documents and manuals.

The potential of the communities to organize and manage their own water resources, has been brought out by many studies (Singh, 1994; Patil, 1994; Fernandes, 2001; Agrawal and Narain, 2002; Ballabh and Thomas, 2002; Hooja, 2002; Raju, 2002). Prior to the canal and other surface irrigation systems, tank irrigation was predominant method of irrigation in ancient India. They were by and large constructed and maintained as commons by the user communities under a system called Kudimaramth (voluntary village labour). Shiva (2002) has described the niceties of Kudimaramath as a system and the process of its decline. According to Shiva, the Compulsory Labour Act of 1858, which mandated farmers to provide labour for repair and maintenance of tanks failed to mobilize community participation as Kudimarmath was originally based on self-management and not forced labour. Sengupta (1993) scientifically documented various kinds of traditional techniques used in irrigation system, most of which are reported to be highly decentralized.

Kadekodi *et al.* (2000) give an account of Kumaons of Uttarakhand managing the water distribution and exercising their water rights quite effectively even today. The traditional *Ahar* system practiced in the Gangetic plains and *Johad* system in Rajasthan are some examples of indigenous water management systems that are highly relevant even today. The experience of Ralegan Siddhi under the leadership of Anna Hazare teaches us that it is possible to break the nexus between dryland and dry life on account of full contribution from the community itself whose life is going to be changed (Kadekodi, 2002).

*Pani Panchayats* as cooperatives based on the concept of managing water resources at the community level on the principle of equal sharing and distribution were initiated in Maharashtra after the 1972 drought that hit the State. Similar experiments are underway in other states like *Piyat Mandalis* (Gujarat), *Pani Panchayats* (Orissa), Water User Associations (Andhra Pradesh), Water User Societies (Karnataka), etc. (Singh *et al.*, 2007).

Deshpande and Reddy (1994) have studied the working of *Pani Panchayats/Water* Councils in Naygoan and neighbouring villages of Purandar taluka of Maharashtra, which is recommended by them provide a framework to model a group dynamics woven around resource sharing under the philosophy of coming together on their own to find solutions to their problems. The success of *Pani Panchayat* were based on the factors of homogeneity (belonging to same region with similar socio-economic conditions), cost effectiveness (cost of lift irrigation, which was less than Rs. 3000), commitment (mandatory collection of 20 per cent share from farmer) and equitable sharing of water.

Andhra Pradesh (AP) government has been a pioneer in introducing reforms in irrigation sector on a large scale with Andhra Pradesh Farmers Management of Irrigation Systems Act, 1997, though the initiation in Participatory Irrigation Management (PIM) began in Gujarat and Maharashtra. AP model of WUAs rests on the models of irrigation reforms introduced in Mexico and Turkey (Peter, 2002). In the AP model the district collector is authorized to delineate every command area for each irrigation system on hydraulic basis and declare them as water users areas for which WUAs should be established. All those who use water and pay water charges are members of WUA. PIM is reported to be successful in AP setting example for other states because of five main factors : (i) political will and support, (ii) legal environment, (iii) supportive role of irrigation development, (iv) financial incentives to sustain the reforms, (v) devolution of powers to users associations (Pangare, 2002).

The Karnataka Irrigation and Certain Law ( Amendment Act, 2000) enables the formation of Water Users Cooperative Societies (WUCSs) to develop irrigation infrastructure, procure water in bulk, operate and maintain canals, levy and collect charges, mobilize resources, create awareness, etc. It is expected that problems of crop violation, illegal utilization and over utilization would reduce with the active participation of WUCs in the near future.

## 3. Watershed Approach to Sustainable Development

Watershed is a basic hydrological unit and an important source of drinking water with a topographical boundary and water outlet comprising of soils, landforms and vegetation. Fresh water use worldwide was about 1,500 km$^3$ yr$^{-1}$ in 1940 and was projected to be 5,000 km$^3$ yr$^{-1}$ in 2000 with increase in the number of water scarce countries from 7 in 1955 to 20 in 1995 and projected increase of 34 in 2025 (Mishra, 2001). The current strategy for rainfed agriculture in India is based on the concept of scientific conservation of rainwater for holistic and integrated development of watersheds and promotion of farming systems approach with a view to evolving models of sustainable agriculture in rainfed areas (Paul, 1998 and Verma and Singh, 2004). A study in Andhra Pradesh showed that installation and administration of small ponds for run-off collection and supplementary irrigation as part of an overall village water resource development project was beneficial (Doherty, 1994). Based on this strategy, a centrally sponsored scheme 'National Watershed Development Project for Rainfed Areas' (NWDPRA) was launched in October, 1990, which has been under implementation during Eighth Plan.

It has been realized that major difficulties to people's involvement in watershed development programs are managerial and technical skills in community mobilization, conflict resolution and institution building based on the study of Doon valley project (Uttarakhand) suggested that the watershed developmental projects should aim at the promotion of watershed development as a viable means of people oriented sustainable development. The mountain areas have specific developmental challenges due to complexity of rural livelihoods and supporting biophysical resources depicted through the adoption of specific farming systems. The human adaptation mechanism to these complexities touching the farming based rural livelihoods. Therefore, on farm watershed development should be adopted as one of the strategic issue. The conventional watershed development approach should deviate towards community watershed whereby food and livelihood needs are satisfied first.

Singh and Kumar (1998) suggest that conservation and better utilization of rainwater is the best strategy in the hilly area since the topography of the region is not suitable for the construction of dams. A government programme initiated in 1981 with the joint efforts from the community is Sukhomajri a village in Himalaya is reported to have succeeded in watershed management. Chopra *et al.* (1990) have studied the functioning of this model of collective action wherein, the common property resources were transferred to village society and the Water User Cooperative Society (WUCS) managed and distributed water on equal rights with active participation of beneficiaries. The "Sukhomajri model" indicates peoples' participation in the management of common property resources in the region with the objective of preservation and self-sustaining development.

**Policy Issue for Sustainable Development**

The action plan for SD should be based on environmental acceptability, technical feasibility and economic viability (Jauhari, 2002). Subsidies, which encourage inefficiencies in resource allocation and sub-optimal use of natural resources should be reduced or removed or rationalized (Gadgil and Guha, 1995; Gulati and Sharma, 1995; Srivastava *et al.*, 2001; Planning Commission, 2000).

To internalize the externalities of pollution, collection of pollution tax from both the producer as well as the consumer is advocated for environmental protection in the spirit of social justice (Chen and Tsai,

2000). Since the command and control strategies have proved to be ineffective instruments of SD, there is a need for the combined role of government, civil society and political actors for SD (Singh, 1998).

Sustainability of natural resources for development can be assured in by regulating the use through administration of various kinds of instruments and initiation of measures for conservation. For example,

- Removal of subsidies,
- Conferring property rights on commons to community and *panchayats,*
- Market intervention for fixing the value of natural resources,
- Economic instruments,
- Replacement of old Acts that were enacted before independence by New Act, also called Indian Forest Act, 1927,
- Population stabilization,
- Ensuring food security to all,
- Meeting housing, water supply, sanitation and health care needs and planning for future needs,
- Clean technologies in production system and development of natural resources,
- Use of renewable energy sources,
- Empowerment of weaker and vulnerable sections,
- Conservation of biodiversity,
- Economic incentives for conservation and efficient use,
- People's participation,
- Institutional reforms and support,
- Technology improvement,
- Mixed role of government, private sector, NGOs and people.

There is felt need for efficient use of water resources and their allocation among competing users in a way that gives priority to the satisfaction of basic human needs and meeting the requirement of preserving the ecosystems and their functions. The new techniques in water resources management are recycling of water, water harvesting, desalination of sea water, private management and participatory management. Since it is unlikely that the institutional intervention can answer major issues of sustainability of natural resources at least in the short-run, the major option for sustainable development of water resources is to create a platform for all the resource users to come together, to discuss and evolve norms for imposing restraint on resource use (Shah, 1993). Sengupta (1993) advocates documentation and adoption of traditional water harvesting systems suiting to local conditions without homogenizing them. According to Biswas (1990) "Efficient use of water is simply not possible, unless it is managed in an environmentally sound manner. Thus, environmentally sound water management is an essential requirement for future development of developing countries, and will become an even more important consideration in the future than it was in the past.

Protection of endangered species, identification and initiatives for hot spot areas, enhancing indigenous and community based biodiversity conservation efforts, promotion of eco-networks and corridors are essential for conservation of biodiversity (World Summit, 2002).

Stabilizing population is an essential requirement for promoting SD with more equitable distribution of alleviation, as the rising population has serious implications on food and water security, health care, rural and urban services and sustainability of ecosystem (Tewari, 2000).

For achieving SD there is need to reduce the pressure on environment by controlling population growth, improving literacy, promoting environmental awareness drives and introducing poverty alleviation programmes. Depéndence on these would reduce the burden on non-removable and renewable but, exhaustive resources. Jodha (2001) opines that "sustainability in agriculture is not possible through traditional measure in fragile regions it requires application of modern science and technology blended with the rationale of indigenous practices". Raghuvanshi and Sajwan (1994) observed that more emphasis should be given to advanced scientific research in different scientific fields through inter-disciplinary plan of action to improve the irrigation management in arid and semi-arid conditions. Alagh and Kashyap (2000) highlighted the role of social activists in water resource conflicts management. Adiseshaiah (1991) proposes that growth centered development plans should be replaced by plans aiming at sustainable development, which will address both to safeguarding and enriching the environment and meeting the basic needs of those who are in need, Muzammil (2001) pinpoints that participatory land management must be made with the consideration that equity in land distribution would help to produce more food.

Major difficulties to people's involvements in watershed development programmes were managerial or technical, and skills in community mobilization, conflict resolution and institution building (Gawande, 2001, Singh and Mahlawat, 2001). Sengupta (1993) calls for learning from the experience of traditional knowledge, "unless they are compared and categorized according to their basic similarities, modern knowledge can not make much headway. Swaminathan (2002) is of the opinion that agriculture in most developing countries is not just a food-producing enterprise, but also the backbone of sustainable ecological and livelihood security systems. In addition, it is for the national sovereignty. Agricultural progress represents the best safety net against hunger and deprivation because of the greatest good it confers on the largest number of people. Therefore, national and international public policies should recognize that food is a powerful tool for socially meaningful development. Swaminathan calls for declaring all good farmland as agri-reserves putting in regulations, which make it difficult to convert prime farmland for non-farm uses without commencing reasons. He suggests agri-reserves and biosphere reserves can become mutually supportive-agri-reserves being the guardians of food security and biosphere reserves of ecological security.

Research should focus on the suitable strategies, which would harmonize compulsions of economic development with imperatives of ecological reservations (Nambiar, 1997). Pachauri and Sridharan (1998) identify some areas that need to be addressed by the water policy in India in the years to come: 'The emerging issues in the management of water resources in India are chiefly of identifying and examining alternative institutional and policy arrangements for example in the context of ground water, the real issue is not over-exploitation but management of ground water, the real issue is not over-exploitation but management of ground water to address a wide array of environmental equity, and sustainability concerns. On the one hand, there is a need to identify local institutional arrangements for the management of water resources by users themselves that ensure equitable and sustainable use of water and on the other, we need to look at the role of improved technology in augmenting the carrying capacity of water resources.

For SD we need to rely on renewable as well as non-exhaustible resource such as solar energy and wind in areas and sectors wherever it is feasible to tap and use them. The consumption of these resources by one does not reduce the magnitude of resource flow to others using or non-using these resources. This would reduce the burden on non-renewable and renewable but, exhaustive resources.

Rai and Yadav (2002) specify incentives and regulatory policies to compensate for externalities related to natural resources, adjustments/ changes, e.g., diversification, crop rotations in crop plans to be facilitated in order to achieve a balanced crop-mix, effective enforcement of procedures for review and approval of the safety of existing and new agricultural chemicals and other agents used in agricultural production, information about sustainable agricultural practices and new policies to encourage wider adoption to be disseminated to farmers to strengthen the cause of 'sustainable agriculture'.

The industries that produce solid wastes need to be comprehensively identified and brought under pollution control laws. The existing laws are weak and not strictly enforced as a result of which there is mismanagement leading to serious impacts on the environment.

According to the TERI report, it is possible to delink the increase in economic activity from pollution. This requires an effective environmental management plan, which include environmental strategy, regulation, institutional capacity - building, and economic incentives and penalties. Besides regulatory and economic tools, such support measures as training and education for the industry, governmental agencies, and the public, as well as greater coordination among institutions, are also important for implementing the plank to put India on the road to sustainable development. In country like India, which faces severe resource constraints, a community-based approach appears to be only viable option for improving water quality and better environmental management and conservation of natural resources (Pachauri and Sridharan, 1998).

However, we fined that there is need for further investigation in many virgin areas, which are essential for developing a long-term policy framework as well as to address to the current micro-level issues.

- Study fail to identify natural resources and human resources linkages.
- Documentation of the state of the displaced households affected by development projects.
- In India and elsewhere, Environmental Impact Assessment (EIA) studies are taken up before a project is initiated. These studies indicate the likely impact. The real impacts occur after project is implemented. The Ministry of Environment & Forests should consider EIA studies before and after implementation of projects in beneficiary and affected areas for understanding the status of affected persons and to review the sustainability of the projects.

- There is absence of studies on the behavioural and economic parameters determining water demand, and water assessments as a part of land and climate or agro-climatic regimes (Alagh, 2003).

Impact assessment of watershed management projects is required in view of the objectives of sustained and stabilized employment, poverty reduction (Nair and Chattopadhyay, 2001 and Singh *et al.*, 2007).

**Research Gaps and Needs**

The research studies have brought in the following issues to limelight:

- The negative role of government in exploiting natural resources for the benefit of corporate sector, rich and urban sector,
- Role of PRIs in natural resource management and SD,
- Relevance of participatory management of natural resources for sustainable development,
- Need for development and conservation of natural resource as support to livelihoods of poor and for sustainable development,.
- Need for continued role of government in protecting and managing natural resources and environment through regulations,
- Role of NGOs in catalyzing the movement and actions for protection of environment for people's well being and maintenance of ecology,
- Lack of environmental modeling in planning,
- Public - private partnership needed, and
- Developing sound database and information system.

Sustainable development along with sustainable agriculture is necessary for the execution of an integrated approach to increase the food production and enhancement of food security and food safety in an environmentally sustainable way. In fact both the concepts are interdependent and interrelated. To create the conditions for sustainable agriculture major adjustments are needed in agricultural, environmental and macro-economic policy, at both national and regional levels. The priority must be on maintaining and improving the capacity of the higher potential agricultural lands to support an exploding population. The main tools of sustainable agriculture and policy aim at agrarian reform, soil conservation and improved management of inputs.

Katyal *et al.* (1998) identified research gaps for SD in agriculture as follows :

- Research on an integrated policy framework on population management and sustainable growth to address to the problems of pressure on land from man and animals,
- Research and development activities on developing resource saving technologies and efficient use, and
- Integrated nutrient management harmonizing chemical and natural nutrient sources in agriculture.

In India regeneration programmes of watershed and regeneration of CPRs has been undertaken on a selective basis, which act as demonstrative models but, given the extent of wastelands and the CPRs, there is a need for systematic study in the manner in which a programme of regeneration of these lands can be undertaken at the national scale (Kadekodi and Perwaiz, 1997). In addition, Prof. Amartya Sen, Nobel Laureate in Economics, proposes the need to reduce poverty and inequality by efficient distribution of the available resources rather than relying on supply.

**Suggested Measures for Sustainable use of Natural Resources**

- To internalize the external costs of exploitation of natural resources through proper pricing of the resources.
- Management of natural resources by community based institutions through discharge of incentives, penalties, contribution, fixation of shares, user charges, etc.
- Environmental protection is intra-sectoral and trans-boundary issue, hence co-operation from public, different sectors, national government and international co-operation, particularly from the neighboring countries is essential for framing a sound environmental policy.
- There is need for decentralization in decision-making relating to local environmental problems, resource use and management.

## CONCLUSION

Many critical survival issues are related to uneven development, poverty and pollution growth. They all place unprecedented pressures on the planet's lands, waters, forests and other natural resources, not least in the developing countries. It is observed that most of the natural resources in India are over exploited like other countries of the world. In India, common property resources (CPRs) are degraded and their productivity is much below the potential that has been demonstrated under good management of the CPRs suffer from what Garredin calls "The Tragedy of Commons". This is the reason that a debate is going on in respect of application of effective tools to manage natural resources to achieve sustainable development. There is an urgent need for their restoration so that they can contribute fully to economic growth and development and help alleviate the problems of poverty, unemployment and ecological degradation. Alternatives to the kind of development design in India are still being searched and sustainable development is one of them. There have been some fundamental problems endemic to sustainable development like transfer of environmental technology, the need for a social policy, etc. But sustainable development appears to be a better alternative in the present context as India has to decide its future in a development policy, which will encapsulate equity, equality and excellence. There is need for restructuring of our priorities to achieve the objective of economic growth and equitable distribution of wealth. It is possible to attain the development goals without causing much damage to the natural resource base.

To conclude, there is a realization that the present generation has a right to enjoy the natural resources and it must be passed on the posterity without any impairment to these resources.

## REFERENCES

Adisheshiah, M. (1991). Natural resource environment of India. *In: The Emerging Challenges.* Sage Publications, New Delhi, pp. 194-234.

Agarawal, A, and Narain, S. (2002). Community and household water management : The key to environmental regeneration and poverty alleviation. *In : Institutionalizing Common Pool Resources* (Ed. Dinesh K. Marothia ). Concept Publishing Company, New Delhi, pp. 115-151.

Agarawal, A, Narain, S. and Sen, Srabani (1999). *The Citizen's Fifth Report, Part 1: National Overview.* Centre for Science and Environment, New Delhi.

Agrawal, V.P. and Rana, S.V.S (1985). *Environment and Natural Resources.* Jagmader Book Agency, New Delhi.

Alagh, Y. K. and Kashyap, S. P. (2000). National Seminar on "Issues Related to Water Resource Use in India" : Role of Social Scientists in conflict management. *In : Proc. of the Seminar on Issues Related to Water Resource Use in India* (Ed. Rohit D. Desai), pp. 15-19.

Alan, Durning (1989). Action at the Grassroots : Fighting Poverty and Environmental Decline. World Watch Paper 88, January, 1989.

American Society of Agronomy (1989). Decision reached on sustainable agriculture. Agronomy News, Madison, Wisconsin, p. 15.

Ballabh, V. and Thomas, P. (2002). Integrated Management of Water and Land for Sustainable Development in Semi-Arid Areas. *In : Institutionalizing Common Pool Resources.* Concept Publishing Company, New Delhi, pp. 152-169.

Barbier, E. (1987). The Concept of Sustainable Development. *Environmental Conservation,* 14 (2) : 101-110.

Bengeri, K. V. (1995). *Outline of Environmental Audit.* Srikrishna Publications, Hubli.

Biswas, A. K. (1990). Objectives and concepts of environmentally-sound water management. *In : Environmentally-sound Water Management* (Eds. N. C. Thanh and A. K. Biswas). Oxford University Press, Mumbai.

Chaudhary, M. S., Tak, T. M. and Sharma, M. L. (1994). Socio-economic and environmental dimensions of irrigation water Management. *In: Socio-economic Dimension of Irrigation* (Ed. R. K. Gurjar). Printwell, Jaipur, pp. 1-12.

Chen, Miao-sheng and Tsai, Ming-Liang (2000). A theoretical model of sustainable development. *Indian J. Economics,* 81 (320) : 12-29.

Chopra, K., Kadekodi, G. K. and Murthy, M. N. (1990). Studies in economic development and planning. *In : Participatory Development and Common Property Resources.* Sage publication, New Delhi, p. 52.

Clarke, Tim (1991). EIA and Sustainable Development. EIA Trainers Newsletter, 6 Center University of Manchester, UK.

Common, M. (1996). *Environmental and Resource Economics - An Introduction,* 2nd ed . Longman, New York.

Common, Michael and Perrings, C. (1992). Toward an ecological economics of sustainability. *Ecological Economics,* 6 (1): 7-34.

Dalal-clayton, Barry (1992). Modified EIA and indicators of sustainability : First step towards sustainability analysis. World Bank Conference on Environment and Sustainable Development, Washington, D.C.

Daly, H. E. and Cobb, J. (1989). *For the Common Good.* Beacon Press, Boston, USA.

Dayanandan, R. (2005). Imperatives of sustainable development in rural India. *In : Sustainable Development : Opportunities and Challenges* (Ed. R. Dayanandan). Serials Publications, New Delhi, pp. ix-xxiv.

Dhanagare, D. N. (2000). Joint Forest Management. : A Study of Uttar Pradesh. G. B. Pant Social Science Institute, Allahabad.

Doherty, V., Mirinda, M. S. and Kampen, J. (1994). Social organization and small watershed development. *In : socio-Economic Dimension of Irrigation* (Ed. R. K. Gurjar). Printwell, Jaipur, pp. 126-150.

El Serafy, S. (1989). The proper calculation of income from depletable natural resources in environmental accounting for sustainable development. World Bank, UNEP Symposium, Washington, D. C.

Fernandes, W. and Asif, M. (1997). Development Induced Displacement in Orissa 1951-1955. Indian Institute of Social Sciences, New Delhi.

Fernandes, W. (1996). Tribal, forests, displacement and sustainable development. *In : Sustainable Development- Ecological and Socio-cultural Dimensions* (Ed. K. Gopal Iyer). Vikas Publishing House Pvt. Ltd., New Delhi.

Gawande, S. P. (2001). Community participation in integrated watershed management for sustainable productivity. *In : Community Participation in Natural Resource management*. Rawat publications, New Delhi, pp. 109-118.

GOI (2004). *India 2004 : A Reference Manual*. Publications Division, Ministry of Information & Broadcasting, Govt. of India, New Delhi.

Goodland, R. and Ledac, G. (1987). Neo-classical economics and principle of sustainable development. *Ecological Modelling,* 38 : 19-40.

Goodland, R., Daily, H. E. and El Serafy, S. (Eds.) [1991]. Environmentally Economic Development Building on Brudtland. Washington, DC.

Govt. of India (1996). *Agricultural Situation in India.* Ministry of Agriculture, GOI, New Delhi.

Heaton, G., Repetto, R. and Sobin, R. (1991). *Transforming Technology : An Agenda for Environmentally Sustainable Growth in the 21st century.* World Resources Institute, Washington, D.C., p. viii.

Holmberg, J. (1991). Indicators of Sustainability. Internal Memorandum, IIED, London, UK.

Hooja, R. (2002). Participatory irrigation management in the Indian context. *In : Users in Water Management* - The Andhra Model and its Reliability (Eds. R. Hooja, G. Pangare and K. V. Raju). Rawat Publications, New Delhi, pp. 3-28.

Human Development Report (1991). United Nations Development Programme (UNDP). Oxford University Press, New York.

IUCN (1980). World Conservation Strategy : Living Resource Conservation for Sustainable Development. IUCN/ UNEP/ WWF, Gland, Switzerland.

Jauhari, V. P. (2002). *Sustainable Development of Water Resources*. Mittal Publication, New Delhi.

Jodha, N. S. (2001). *Life on the Edge : Sustaining Agriculture and Community Resources in Fragile Environments*. Oxford University Press, New Delhi.

Kadekodi, G. K. (2002). Towards an Alternative Approach of Rural Development : A case Study of Ralegaon - Siddhi. CMDR, Dharwad in collaboration with FRCH, Pune.

Kadekodi, G. K. (2004). *Common Property Resource Management: Reflections on Theory and the Indian Experience*. Oxford University Press, New Delhi.

Kadekodi, G. K. and Perwaiz, A. (1997). Dimensions of wastelands and common property resources in India. *In : Environmental Database of Indian Economy* (Eds. J. K. Parikh and V. K. Sharma). Indira Gandhi Institute of Development Research, Mumbai, pp. 63-86.

Kadekodi, G. K., Murthy, K. S. R. and Kumar, K. (2002). *Water in Kumaon: Ecology, Value and Rights*. Gyanodaya Publications, Nainital.

Katyal, J. C., Ramachandran, K. and Reddy, M. N. (1998). Ecological Agriculture and Sustainable development . Indian Ecological Society and Center for Research in Rural and Industrial Development, Chandigarh.

Kohli, Anju, Shahi, Farida and Choudhary, A. P. (1997). *Sustainable Development in Tribal and Backward Areas*. Indus Publishing Company, New Delhi, pp. 1-84.

Lockeretz, W. (1988). Environmentally sound agriculture. *American J, Alternative Agriculture,* 3 : 174.

Mahapatra, L. K. (1994). *Tribal Development in India : Myth and Reality*. Vikas Publications, New Delhi.

Manikandan, S. and Rao, K. Raja Mohan (2005). Common property resources and sustainable development : A perspective. Chapter 5, *In : Sustainable Development : Opportunities and Challenges* (Ed. R. Dayanandan ). Serials Publications, New Delhi, pp. 45-53.

Mathew, J. T. and Tunstall, D. B. (1991). Moving towards eco-development: Generating environmental information for decision-makers, Issues & Ideas paper. World Resource Institute, Washington, DC.

Mishra, A. (2001). *Watershed Management*. Author Press, New Delhi.

Mishra, G. P. and Bajpai, B. K. (Eds.) [2001]. *Community Participation in Natural Resource Management.* Rawat Publications, Jaipur, pp. 13-34.

MOEF (2001). *India : State of Environment.* Govt. of India, New Delhi.

MOEF (2005). Empowering people for Sustainable Development. Government of India (http://envfor.nic.in /divisions/ ic/ wssd/doc 1/home, htm accessed on 8. 8. 2005).

Mohanty, Tapan, R. (1999). Sustainable Development : An overview. *Employment News,* 24 (5) : 1-2.

Moolya, A. (1998). A Note on Environmentalism. Presented is the Seminar during Refresher Course in Economics on May 23, 1998 at Mangalore University, Mangalagangotri, Konaje, Karnataka.

Morvaridi, B. (1997). The environmental impact of irrigation : The Social dimension (A Case Study of Sultanpur India), *In : Sustainable Development in a Developing World* (Eds. Kirkpartrick Colin and Lee, Normah). Deward Elgar Publishing Ltd., UK, pp. 231-244.

Murthy, M. N. (1994). Management of common property resources : Limits to voluntary collective action. *Environmental and Resource Economics,* 4(6) : 581-594.

Muzammil, M. (2001). Liberalisation and land Management, *In : Community Participation in Natural Resource Management* (Ed. G. P. Mishra and B. K. Bajaj). Rawat Publications, New Delhi, pp. 134-140.

Nair, K. N. and Chattopadhyay, Srikumar (2001). Issues and perspectives in Watershed Management: A report. *In : Watershed Management for Sustainable Development : Field Experiences and Issue* (Eds. K. N. Nair and Srikumar Chattopadhayay). Center for Development Studies, Thiruvanthapuram, Kerala.

Nambiar, K. V. (1997). Conservation of biodiversity : A major policy imperative for environmental planning and Sustainable development. *In : Conservation and Economic Evaluation of Biodiversity,* Vol. II (Eds. P. Pushpangadan, K. Ravi and V. Santhosh). Oxford & IBH Publishing Co. Pvt. Ltd., New Delhi, pp. 461-464.

O' Meara, M. (1999). Exploring a new vision for cities. *In : State of the World 1999* (Eds. Brown *et al.).* Earthscan/World Watch Institute, London, p. 136.

OECD (1991). *The Economics of Sustainable Development : A Progress Report.* Organisation for Economic Cooperation and Development, Paris, France.

Pachauri, R. K. and Sridharan, P. V. (1998). *Looking Back To Think Ahead: Green India 2047.* Tata Energy Research Institute (TERI), New Delhi.

Pangare, G. (2002). Scaling up participatory irrigation management in India: Lessons from the Andhra Pradesh model and strategies. *In : Accounting and Valuation of Environment* : Case Studies from the ESCAP Region (Eds. R. Hooja, G. Pangare, J. K. Parikh and K. S. Parikh). United Nations, New York.

Parikh, K. S. (1989). An operational and measurable definition of sustainable development. Indira Gandhi Institute of Development & Research, Mumbai.

Patil, R. K. (1994). Economics of farmer's participation in irrigation management. *In : Socio-Economic Dimension of Irrigation* (Ed. R. K. Gurjar). Printwell, Jaipur, pp. 186-199.

Pearce, David W. and Warford, Jeremy J. (1993). World Without End-Economics, Environment and Sustainable Development. Oxford University Press, New Delhi.

Poffenberger, M. and Mc Gean, Besty (Ed.) [1996]. *Village Voices, Forest Choices : Joint Forest Management in India.* Oxford University Press, New Delhi.

Raghuvanshi, C. S. and Sajwan, K. S. (1994). Strategies for water management in arid and semi-arid : A critique. *In : Planning and Policies of Irrigation Management.* (Ed. R. K. Gurjar). Printwell, Jaipur.

Rai, K. N. and Yadav. D. B. (2002). Sustainable development of Agriculture in Haryana (India). *In : Sustainable Agriculture Poverty and Food Security,* Vol. II (Ed. S. S. Acharya). Rawat Publications, New Delhi, pp. 903-917.

Raju, K. V. (2002). Participatory irrigation management in Andhra Pradesh: The way forward. *In : Users in Water Management - The Andhra Model and its Replicable* (Eds. R. Hooja, G. Pangare and K. V. Raju). Rawat Publications, New Delhi, pp. 83-126.

Rana, S.V.S. (2005). Global Environmental Problems. Chapter 22, *In: Essentials of Ecology and Environmental Science.* Prentice Hall of India Pvt. Ltd., New Delhi, pp. 384-430.

Rau, J. E. and Wooten, D. C. (1985). *Environmental Impact Analysis Handbook.* McGraw Hill Publishing Company, Columbus, USA.

Raymond, F. Mikasell (1992). Environmental Assessments and Sustainability Project and Programme Level. *In : World Bank Conference on Environment and Sustainable Development,* Washington, DC.

Reddy, Bhaskar (1995). Shifts in crops, cropping systems, diversification and competitive advantage for Andhra Pradesh. Special Lecture Delivered at 31[st] REAC Meeting, ANGRAU, Rajendra Nagar, Hyderabad.

Reddy, V. Ratna (1995). Environment and sustainable development: Conflicts and contradictions. *Economic & Political Weekly,* 30 (20): A 21.

Repetto, R. (1988). The Forest for the Trees ? Government Policies and the Misuse of Resources. World Resource Institute, Washington, D C.

Repetto, R. (Eds.) [1985]. *The Global Possible.* Yale University Press, Bedford, London.

Ryding Seven, Olof (1992). *Environmental Management Handbook.* Lewis Publishers, Florida, USA.

Sahu, G. C., Mishra, K. N. and Mishra, A. (2002). Natural resources of coastal region of Orissa. *In : Natural Resources of Coastal Regions of India.* Publication of TNAU and NBSS & LUP, pp. 90-112.

Sahu, N. C., Nayak, B., Maharana, D. P. and Nayak, B. (2000). Economic valuation of environment : Components and complexities. *In : Environment and Economic Development* (Ed. N. Rajalakshmi). Manak Publications Pvt. Ltd., New Delhi, pp. 3-31.

Sehgàl, J. and Abrol, I. P. (1994). *Soil Degradation in India : Status and Impact.* Oxford & IBH Publishing Co. Pvt. Ltd., New Delhi, p. 80.

Sehgal, S. (2001). Joint Forest Management : A Decade and Beyond, A Workshop Paper. Institute of Economic Growth, New Delhi.

Sen, Amartya (1981). *Poverty and Famine.* Oxford University Press, New Delhi.

Sengupta, N. (1993). *User-Friendly Irrigation Designs.* Sage Publications, New Delhi.

Senthi, R. M. (1996). Green revolution and environmental issues : Action programmes for sustainable development. *In : Sustainable Development-Ecological and Social Dimensions* (Ed. K. G. lyer). Vikas Publishing House Pvt. Ltd., New Delhi.

Shiva, Vandana (2002). *Water Wars : Privatization, Pollution and Profit.* Pulto Press, London.

Simon, J. and Khain, H. (Eds.) [1984]. *The Resourceful Earth.* Basil Blackwell, Oxford, UK.

Singh, K. K. (2008). *Natural Resources Conservation & Management.* MD Publications Pvt. Ltd., New Delhi, pp. 1-11.

Singh, K. K. and Verma, Gayatri (2000). Agrochemicals : Virtue or Bane. *Science & Culture, 69* (9-10): 296-298.

Singh, K. K. (1994). Irrigation management by farmers : The Indian Experience. *In : Socio-Economic Dimension of Irrigation* (Ed. R. K. Gurjar). Printwell, Jaipur, pp. 158-185.

Singh, K. K., Tomar, Alka, Singh, Vinod and Singh, Mahadevi (2007). Watershed development for land and water management in arid ecosystem. Chapter 5, *In : Environmental Degradation and Protection* (Eds. K.K. Singh *et al.*), Vol. I. M.D. Publications Pvt. Ltd., New Delhi, pp. 144-187.

Singh, M. (1998). Towards Sustainable Development. *Man and Development,* 20(1): 28-32.

Singh, M. and Lal, M. (2001). Participatory management of natural resources. *In : Community Participation in Natural Resource Management* (Eds. G. P. Mishra, and B. K. Bajpai). Rawat Publications, Jaipur.

Singh, Mahadevi, Choudhary, Anjana and Tomar, Alka (2007). Water councils/*pani Panchayats* for equitable sharing of water. Chapter 36, *In : Water Resources Management and Development* (Eds. K. K. Singh *et al.).* Kalyani Publishers, Ludhiana, pp. 567-585.

Singh, R. and Kumar, R. (1998). Rain water management : An essential component for sustainability in hill agriculture. *In : Ecological Agriculture and Sustainable Development,* Vol. I (Ed. G. S. Dhaliwal). Indian Ecological Society, Ludhiana and Center for Research in Rural and Industrial Development, Chandigarh, pp. 173-180.

Singh, Shekhar (1989). The global environmental debate. *The Indian J. Public Administration,* 35 (3): 371.

Speth, G.J. (1989). The Global Environmental Agenda. *Environmental Science and Technology,* 23(7), July, 1989.

Srivastava, D. K., Rao, B., Chakroborty, P. and Rangamannar, T. S. (2001). *Budget Subsidies in India.* National Institute of Public Finance and Policy, New Delhi.

Sunder, N., Mishra, A. and Peter, N. (1996). Defining the Dalki Forest-Joint forest Management in Lapanga. *Economic & Political Weekly,* 31 (45 and 46) : 3021-3025.

Suri, R. K. and Rawat, P. S. (1996). Sustainable development. *In : Mining Environment and Forest* (Ed. R. K. Suri). Society of Forest & Environmental Managers (SOFEM), Dehradun.

Swaminathan, M. S. (1996). *Sustainable Agriculture : Towards Food Security.* Konark Publishers Pvt. Ltd., New Delhi, pp. v-vii.

Swaminathan, M. S. (2002). *From Rio De Janeiro To Johannesburg : Action Today and Not Just Promises for Tomorrow.* East West Books Pvt. Ltd., Chennai.

Swanson, T. and Barbier, E. (2005). The ecology of an industrial planet. Chapter 1, *In : Managing Sustainable Development* (Eds. Michael Carley and Ian Christie). Earthscan Publications Ltd., London, UK, pp. 1-24.

TERI (1999). *The TERI Energy Data Dictionary and Year Book* (TEDDY). TERI, New Delhi.

Tewari, D. N. (2000). Population and Sustainable development. *Yojna,* 44(8) : 5-14.

Turner, R. Kerry (1993). *Sustainable Environmental Economic and Management: Principles and Practice.* Belhaven Press, Landon.

UNCED (1992). The Global Partnership for Environment and Development: A Guide to Agenda 21, Geneva, Switzerland.

Verma, Gayatri and Singh, K.K. (2004). Natural resource management for Sustainable productivity in dryland areas. Chapter 9, *In : Sustainable Resource Management* (Eds. B.B.S. Kapoor and Ahmed Ali). Madhu Publication, Bikaner, pp. 139-162.

WCED(1987). *Our Common Future.* World Commission on Environment and Development. Oxford University Press, New York.

Wheeler, Stephen M. (2004). *Planning for Sustainability-Creating Livable, Equitable and Ecological Communities.* Taylor & Francis Group, Routledge, London.

Winpenny, James T. (1991). *Values for the Environment - A Guide to Economic Appraisal.* HMSO Books for the Overseas Development Institute, London, UK.

World Bank (1987). World Development Report 1987. Oxford University Press, New York, pp. 1-288.

World Bank (1991). *Environmental Assessment Source Book,* Vol. I to III. World Bank, Washington, DC.

World Bank (1999). *Initiating and Sustaining Water Sector Reforms: A Synthesis.* The World Bank, Washington, DC.

World Conservation Union (1991). Caring for the Earth. United Nations Environment Programme (UNEP), Gland, Switzerland.

World Summit (2002). World Summit on Sustainable Development- Plan of Implementation. *Young Indian,* 13 (10) : 18-20.

WRI (1992). Guide to Global Environment, Washington, D C., USA.

WRI (1992). *World Resources 1992-93.* World Resources Institute. Oxford University Press, New York, USA.

## Chapter 9

# Global Warming, Ozone Layer Depletion and Climatic Change: A Global Perspective

*K.K. Singh[1], Ali Akram[2], A.K. Singh[3], Vijay Pal Singh[4] and Vinod Singh[5]*

## INTRODUCTION

The Earth's climate has changed throughout history and is still in a state of constant turmoil. From glacial periods or "ice ages", where ice covered significant portions of the Earth, to interglacial periods where ice retreated to the poles or melted entirely - the climate has continuously changed. The Earth is only 5 to 9°F, or about 3-6°C, warmer today than it was 1 0,000 years ago during the last ice age. Weather refers to the day-to-day state of atmosphere, while the climatic variability refers to year-to-year and decade to decade patterns of weather. Thus, the climatic change refers to long-term trends. Climatic changes have occurred on the Earth due to a variety of factors like evolution of the chemical composition of the atmosphere, geological changes in the earth, slow changes in the Sun-Earth geometry etc. Many such climatic changes are related to natural causes and man has no control over them, e.g., the cycle of glacial cold

[1]Project Directorate (Research), Agriculture & Soil Survey, Krishi Bhawan, Bikaner (Raj.) - 334001.

[2]Office of the DEO-DPC, Sarva Shiksha Abhiyan, Bikaner (Raj.) - 334001, India.

[3]Department of Physics, Dronacharya College of Engineering, Farrukhnagar, Gurgaon (Har.) - 122001, India.

[4]Department of Agricultural Zoology & Entomology, RBS College, Bichpuri, Agra (U.P.) - 283105, India.

[5]Department of Geography, Govt. Dungar College, Bikaner (Raj.) - 334001, India.

and interglacial warm epochs on the Earth that have been inferred through proxy climatic records, have resulted in catastrophic changes in the Earth's climate (Singh, 2008).

The global warming and ozone layer depletion, due to increased anthropogenic emission of green house gases in the atmosphere, have drawn worldwide attention with an alarming stage of iceberg melting; rise in sea level; upsets in local and global ecosystems; changes in rainfall patterns; depleting forests; increased pathogenesis in plants, animals and human beings; extinction of large number of plant and animal species; drought and desertification; soil erosion and land degradation leading to alteration in living style of the people due to the above changes. The environmental degradation threatens the very survival of human beings. The world's leading scientists project that during our children's lifetimes global warming will raise the average temperature of the planet by 2-6°F, or 1-3.5°C (Tandon *et al.,* 2005).

Several reports of the United Nations Inter-governmental Panels on Climatic Changes (IPPC, 1990,1994, and 1996) indicated the urgency of the problem. IPPC (2001) has warned that by the middle of this century, the globe's temperature will rise by anything upto 5.8°C. The burgeoning human population at an annual growth rate of two per cent, with an increased rate of energy consumption at the rate of four per cent, and rapid industrialization are the major causes of increasing atmospheric concentration of green house gases leading to global warming. The paper attempts to discuss various causes of climatic change, global warming and its impact on human lives.

## ENERGY CONSUMPTION

Fossil fuels, chiefly coal, oil and natural gas, now supply most of the world's energy. Only a small amount comes from renewable energy sources, which do not release gases that trap heat in the atmosphere. If we could get more of our energy from renewable sources, we could reduce the amount of fossil fuels we burn. By the year 2050, renewable sources could provide 40% of the energy needed in the world. Use of renewable energy can help both to slow global warming and to reduce air pollution (Fig. 1 & 2).

These fossil fuels, coal, oil and natural gas, emit greenhouse gases when burned. Coal emits high amounts of greenhouse gases, and the world may be supplied with enough of it to last over 100 years. Oil emits high amounts of greenhouse gases and also other types of air pollutants, harmful to the environment. Its supply is also estimated to last over 100

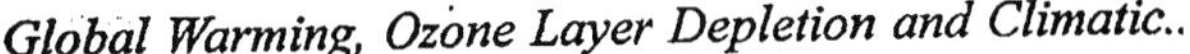

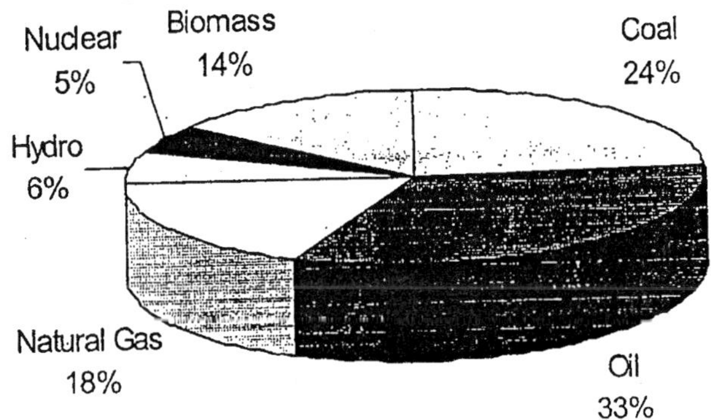


**Fig. 1: The Present Ways of Producing Energy**

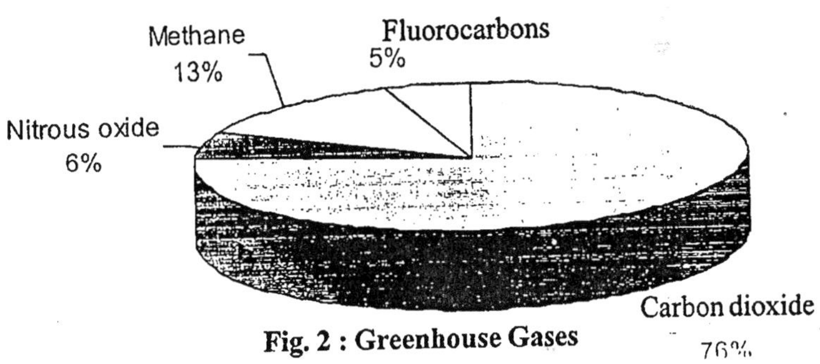


**Fig. 2 : Greenhouse Gases**

years. Natural gas is the lowest of all fossil fuels in greenhouse gas emissions and supplies of that are projected to last over 100 years.

The green house gas emissions in 2000 from energy sector are power (24%), industry (14%), transport (14%), and buildings (8%). Non-energy emissions are $CO_2$-landuse (18%) and non-$CO_2$ agriculture (14%) and waste (3%).

## EARTH'SATMOSPHERE AND GREEN HOUSE EFFECT

The Earth's atmosphere consists mainly of four layers, viz., troposphere, (lower), stratosphere, mesosphere and thermosphere (upper one). The troposphere, in unpolluted state, consists of $N_2$ (78.08%), $O_2$ (20.95%) and traces of other gases such as $CO_2$ (0.0325%), $O_3$ (0.000002%), argon (0.93%), neon, helium, krypton etc.; with varying quantities of water of 1-4% by volume. The major atmospheric constituents ($N_2$ and $O_2$) are not green house gases. This is because homonuclear diatomic molecules such as $N_2$ and $O_2$ neither absorb nor emit infrared radiation, as there is no net change in the dipole moment of these molecules. The stratosphere is above the troposphere and begins at an average height of 12-15 km above the earth. It differs from the troposphere in that it contains very little water vapours and no clouds. Though its gaseous composition is similar, the gaseous mass is only 15 per cent of the atmosphere and it contains more ozone ($O_3$, an isotope of oxygen). The ozone layer absorbs the shortwave UV radiation (below 0.3m) of the sunlight that passes through it. This property gives it an important function in protecting the biosphere. If the ozone layer were not present, then the ionizing UV rays would destroy most or all of the life on earth.

The greenhouse effect (Gh effect) was first described by the French mathematician J. Fourier in 1829. It is also called the global warming or carbon dioxide problem; now extended to chlorofluorocarbon (CFCs), hydrochlorofluorocarbons (HCFCs), methane, nitrogen oxide, etc. The Gh effect or blanket effect is a phenomenon based on well established principles of infra-red (IR) absorption characteristics of gases.

As energy from the sun passes through the atmosphere a number of things take place. A portion of energy (26% globally) is reflected or scattered back to space by clouds and other atmospheric particles. About 19% of the energy available is absorbed by clouds, gases (like ozone) and particles in the atmosphere. Of the remaining 55% of the solar energy passing through the Earth's atmosphere, 4% is reflected from the surface back to space. On an average, about 51% of the sun's radiation reaches the surface. This energy is then used in a number of processes, including the heating of ground surface; the melting of ice and snow and the evaporation of water; and plant photosynthesis.

The greenhouse effect is a naturally occurring process, which aids in heating the earth's surface and atmosphere that is required in sustaining life on the planet. It results from the fact that certain atmospheric gases, such as carbon dioxide, water vapour, and methane are able to absorb longwave radiation emitted from the heated up Earth's surface. Without the greenhouse effect, life on this planet would probably not exist, as the average temperature of the earth would be a chilly -18°C, rather than the present 15°C (Masters, 1994). Greenhouse effect, thus, is essential for keeping the globe/earth in the condition congenial for sustaining life.

There have been numerous warnings in recent decades that the Earth's climatic patterns - which followed an almost steady course for centuries - are undergoing visible changes. This is being caused by changes in the composition of global atmosphere, which consists of a number of natural or synthetic gases. Any increase in the amount of certain gases - particularly synthetic gases - results in the absorption of infrared terrestrial radiations reflected from the surface of the earth. This leads to an enhanced heat-trapping capacity of the earth and atmosphere due to increased concentrations of green house gases (GHGs) which lead to the enhancement of natural green house effect, resulting into universal increase in temperature, causing global climate change (Fig. 3).

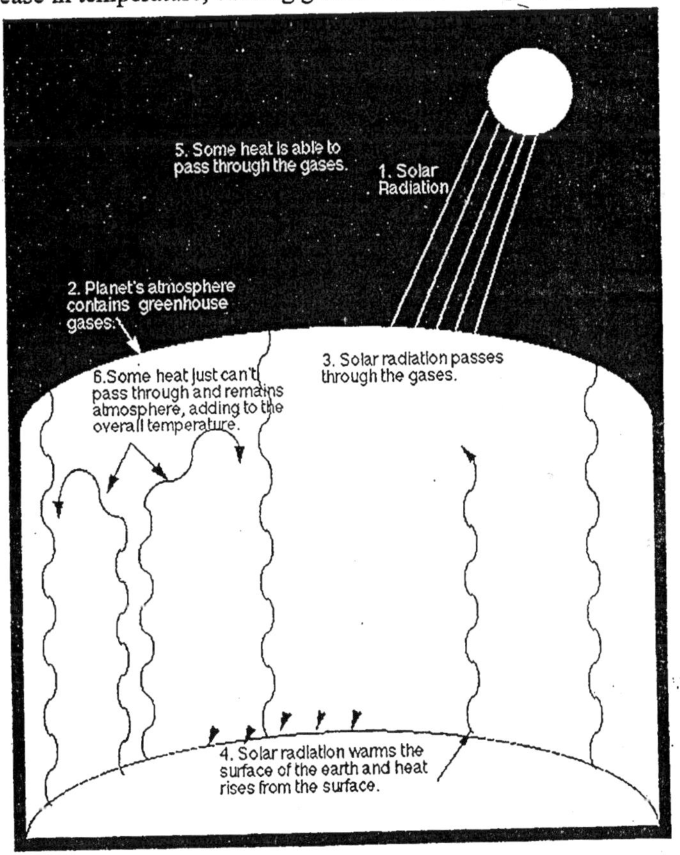


**Fig. 3 : Global warming and greenhouse effect**

A number of gases are involved in the human-caused enhancement of the greenhouse effect (Table 1). These gases include : carbon dioxide ($CO_2$), methane ($CH_4$), nitrous oxide ($N_2O$), chlorofluorocarbons ($CF_xCl_x$), and tropospheric ozone ($O_3$). Of these gases, the single most important gas is $CO_2$ which accounts for about 55% of the green house effect. The

**Table 1: Gases involved in the greenhouse effect**

| *Gas* | *Concentration 2003* | *Greenhouse Contribution (%)* | *Rates of increase (% per year)* | *Natural and Anthropogenic Sources* |
|---|---|---|---|---|
| Carbon dioxide($CO_2$) | 376 ppm | 60 | 0.4-0.5 | Burning fossil fuel & biomass, organic decay, forest fires, volcanic eruptions, deforestation, land-use change, cement making, breathing of animals and humans |
| Methane ($CH_4$) | 1.79 ppm | 16-18 | 0.7-0.9 | Wetlands, termites, natural gas and oil extraction, gas pipe leaks, coal, rice cultivation, cattle, refuse landfills |
| Nitrous oxide ($N_2O$) | 319ppb | 5 | 0.20 | Burning fossil fuels and forests, grasslands, oceans and fertilizers |
| Ozone ($O_3$) | unknown | 8 | 0.25 | Created naturally by the action of sunlight on molecular oxygen and artificially through photo-chemical smog production |
| Chlorofluoro - carbons ($CFC1_3$ and $CF_2Cl_2$) | 880 ppt | 4-8 | 0-4 | Coolant in refrigerators, AC, aerosol spray propellants, cleaning solvents |

**Source** : White and McGovern, 1993 and Khan *et al.*, 2001 contributions of the other gases are 25% for chlorofluorocarbons, 15% for methane and 5% for nitrous oxide. Nitrogen oxide is emitted into the atmosphere from the nitrogenous fertilizer application to the soil, fossil fuel and biomass burning. Greenhouse effect of CFCs and HCFCs is 12000 to 15000 times more potent than $CO_2$ on a molar basis, and due to slow reacting nature these are also powerful agents for destruction of the protective ozone layer in the stratosphere (Encyclopaedia Britannica, 1988).

## Global Warming

Today man-made global warming is occurring much faster than it had at any other time in at least last 10,000 years. Scientists point out the fact that 75 million years ago, the earth's average temperature was 10°C higher than it is today. Conditions were warmer and more humid, but life sustained. Another phenomenon to be taken into account is the "little ice age", which occurred from 1550-1850 A.D. Conditions around the world were cooler than usual, many bodies of water froze over. Since industrial revolution, anthropogenic activities have had an increasing influence on the global climate, a change usually referred to as Global Warming. The average global temperature since the little ice age has risen by 1°F (Singh, 2008).

A marginal rise in atmosphere temperature due to climatic change may seriously upset the ecological balance. Warmer temperatures can lead to contention over new ice-free sea-lanes in the arctic or more accessible resources in the Antarctic. What is more threatening is the prospect of sea level rise and the possible submergence of many coastal and other low lying areas, such as Bangladesh and Maldives. Already, estimated global sea level has risen between 10 to 20 cm over the past 100 years and much of this may have come from increase in global mean temperature.

## Ozone Layer Depletion

Ozone is an isotope (allotrope) of oxygen. Ozone molecule is made up of three oxygen atoms and is denoted by symbol $O_3$. It is a strong oxidizing and chemically highly reactive gas and is present freely in atmosphere in varying quantities. About 90 per cent of the total ozone in the atmosphere is located in the stratosphere and only about 1 0 per cent is in the layer upto 12 km from the earth's surface. Concentration of ozone begins to increase from about 12-14 km, and the maximum is achieved at a height of about 25 km in the stratosphere (Singh *et al.*, 2007).

Ozone is produced in the stratosphere by ionization or photodissociation of oxygen in the presence of UV radiation from the Sun. In spite of splitting of oxygen molecules, oxygen remains abundant in the stratosphere by reconversion of ozone into oxygen. The process of ozone production takes place as follows :

$$O_2 \xrightarrow{\text{UV rays }(\lambda 240\text{nm})} O + O$$

$$O + O_2 \longrightarrow O_3$$

$$O_3^* + M \longrightarrow O_3 + M$$

$O_3^*$ denotes an unstable energetic intermediate form of $O_3$. The excess energy in $O_3$. is removed by collision with atmospheric molecules (M).

The ozone depleting substances (ODS) have atmospheric life times long enough to allow them to be transported by wind into the stratosphere. They release chlorine and bromine, when they breakdown and damage protective ozone layer. The series of events leading to the destruction of ozone layer can be summarized as :

$$O_2 + \text{UV radiation} \longrightarrow 2\,O$$

$$2\,O_3 + 2Cl\ (\text{from CFCs}) \longrightarrow 2\,ClO + 2\,O_2$$

$$2\,ClO + 2\,O \longrightarrow 2\,Cl + 2\,O_2$$

The net effect is

$$2O_3 \xrightarrow{\text{CFCs}} 3\,O_2$$

The CFCs, which are used in the refrigeration industry and household cans of aerosol, deodorants, etc., are also agents of destroying ozone because they release free chlorine (Cl) atom and free radical ClO into the stratosphere which destroy ozone. A single chlorine atom can cause destruction of over 100,000 ozone molecules. Other depleting agents are halons and selected solvents such as methyl chloroform, methyl bromide and carbon tetrachloride.

Small quantity of ozone also occurs in the troposphere and near the earth's surface. Pre-industrial era estimate of global average troposphere ozone is 25 dobson units. Small quantities of $O_3$ can also be locally produced by flashes of atmospheric electricity in thunderstorms or through any electric spark.

In pristine environment, surface ozone is present in very minute quantity of about 10-20 ppb molecules by volume (ppbv) but it exceeds even 100 ppbv in highly polluted environments. This is injurious to human health and also to vegetation. Ozone is a greenhouse gas in the troposphere, and hence, Contributes to global warming.

**Carbon Cycle**

The processes of transport and transformation of various substances in the environment, through living organisms, the atmosphere, oceans, land, and ice, are known collectively as biogeochemical cycles. The Earth system is composed of a number of biogeochemical cycles, all powered by the sun's energy. These global cycles include the circulation of certain elements, or nutrients, upon which life and the earth's climate depend. Carbon is one of the most significant elements in that cycle. Plants and animals are approximately 50% (by dry weight) carbon. Carbon in the form of carbon dioxide ($CO_2$), carbon monoxide (CO) and methane ($CH_4$) is also a significant contributor to greenhouse gases that trap the sun's energy, as it is stored and released as long wave terrestrial emissions from the earth's surface.

Carbon represents only 0.27 per cent of the elements in the earth's crust, but because it exists in both reduced and oxidized states it is vital for life processes (Houghton and Skole, 1993). Only plants and some microbes are capable of reducing carbon. Cellulose, carbohydrates, proteins and fats are all forms of organic matter or reduced carbon. Ultimately, they are all derived from photosynthesis, which converts the sun's energy into embodied (or latent) energy. Fossil fuels represent organic matter that was formed millions of years ago that escaped oxidation and was buried in the earth.

Oxidation occurs through two processes, respiration and combustion. All living organisms engage in respiration; it is the process that releases the latent energy stored in organic matter or fossil fuels. Burning of fossil fuels that have been buried for millions of years beneath the earth's surface provides a net contribution of new carbon dioxide to the short-term system (as opposed to the long-term system in which cycles occur in geologic time over millions of years). It is the combustion of fossil fuels that has now raised the spectre of climatic change.

The major reservoirs of carbon are, in decreasing order, the intermediate and deep oceans (36,730 billion tonnes, or Gt), fossil fuels (4,130 Gt), soil (1,200 Gt), atmosphere (720 Gt), surface ocean (670 Gt), and plants and animals (biomass at 600-1,000 Gt) (Falkowski, 2000). Rocks also store some 75 million Gt of carbon, but these are largely inert and, therefore, do not contribute to the carbon cycle. The most significant fluxes occur between the biota/soil layer and the atmosphere (on the order of 120 Gt per year of uptake and release by the biota/soil layer), followed by the ocean surface and atmosphere (to the order of 100 Gt/year in both directions, with a net uptake by oceans of 2.5 Gt) (Fig. 4).

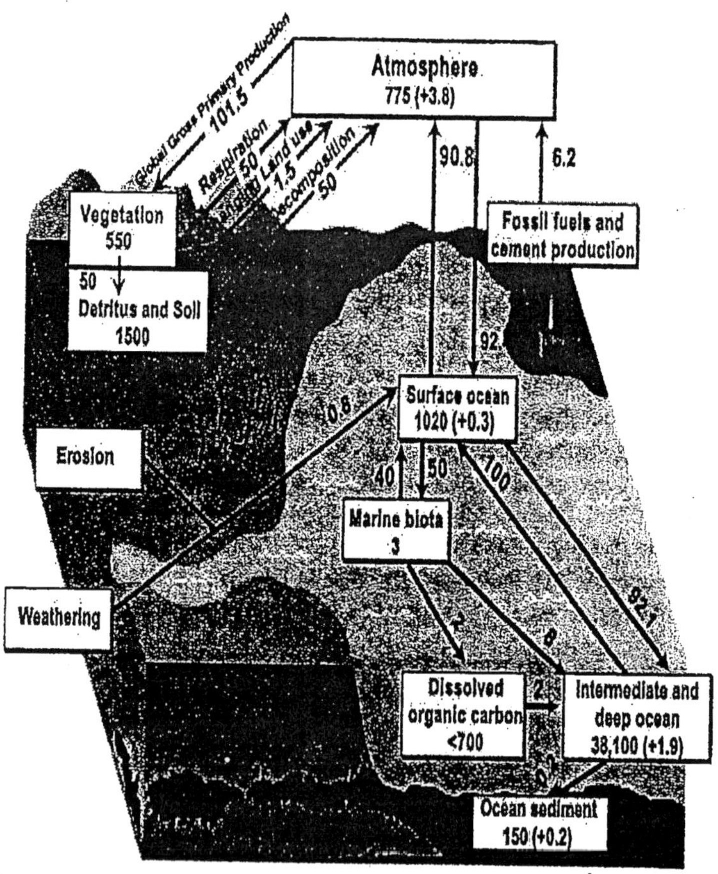


**Fig. 4: The Carbon Cycle**
**(Based on Schimel *et al.*, 1995)**

Throughout most of human history these flows have been in approximate balance. Only recently has the consumption of fossil fuels, which contribute 5.2 Gt/year to the atmosphere, and the large-scale transformation of landscapes, which contributes a net of between 0.4 and 2.5 Gt/year, threatened to upset this balance (Houghton and Skole, 1993). The result is that carbon dioxide concentrations in the atmosphere have risen by 28 per cent since 1850, from approximately 285 ppm to roughly 366 ppm at present (IPCC, 2000). If we take a longer time horizon, concentrations have actually risen by two-thirds, from the 220 ppm average that has been in effect for the past 4,20,000 years (Falkowski, 2000).

The uptake of carbon in so-called "sinks" (reservoirs of carbon) is governed in part by other nutrient cycles. Falkowski (2000) noted that humans have also had a major impact on the cycles of nitrogen, phosphorus and sulphur. Agricultural activities (Chemical fertilization and nitrogen fixing plants) have doubled the fluxes for nitrogen and quadrupled the fluxes for phosphorus. Fossil fuel and biomass burning emissions have doubled the fluxes of sulphur. Though the increase in these other nutrients should in theory increase carbon uptake, the effects of these changes on the carbon cycle are poorly understood.

The uptake and release of carbon by natural and anthropogenically altered ecosystems will also be affected by temperature. According to the latest IPCC scenarios, average global temperatures will increase by between 1.4 and 5.8°C by 2100, and by even more at higher latitudes. One of the "wild cards" of global warming is that scientists do not know precisely how this warming will affect carbon fluxes from biomass and soils. Depending on how ecosystems respond, they may either mitigate or reinforce warming trends through uptake or release of carbon dioxide.

In terrestrial ecological systems, carbon is retained in live biomass, decomposing organic matter, and soil. Changes in land-use can result in the release of carbon into the atmosphere, or withdrawal of carbon from it (WRI, 2000).

### Methane as a Greenhouse Gas

Methane is the second major gas after carbon dioxide responsible for the warming of the environment and ozone depletion. Its presence in the atmosphere was first detected by Migeotte (1948). Its present concentration of 1800 ppbv in the atmosphere is more than double that of about 750 ppbv estimated a hundred years ago (Khalil *et al.*, 1993). Various estimates of global methane production ranged between 350-820 Tg/year (IPPC, 1992; Moss, 1994; Tyler, 1991; Khan *et al.*, 1996 & 2001) (Table 2).

**Table 2 : Estimate of $CH_4$ released into the atmosphere**

| *Source* | *$CH_4$ emission Tg ($x10^{12}$ $gyr^{-1}$)* | *Percent of total* |
|---|---|---|
| **A. Biogenic** | | |
| Ruminants | 80-100 | 12-22 |
| Termites | 25-150 | |
| Paddy fields | 70-120 | |
| Natural wetlands | 120-200 | |
| Landfills | 5-70 | |
| Oceans and lakes | 1-20 | |
| Tundra | 1-5 | |
| Total | 302-665 | 81-86 |
| **B. Abiogenic** | | |
| Coal-mining | 10-35 | |
| Natural gas flaring & venting | 10-30 | |
| Industrial & pipeline losses | 15-45 | |
| Biomass burning | 10-40 | |
| Methane hydrates | 2-4 | |
| Volcanoes | 0.5 | |
| Automobiles | 0.5 | |
| Total abiogenic | 48-155 | 13-19 |
| Grand Total | 350-820 | |

**Source** : Tyler, 1991; IPPC, 1992; Khan *et al.*, 2001

## CAUSES OF CLIMATIC CHANGES

The amount of heat energy added to the atmosphere by the greenhouse effect is controlled by the concentration of greenhouse gases in the Earth's atmosphere. All of the major greenhouse gases have increased in concentration since the beginning of the industrial revolution (about 1700 AD). As a result of these higher concentrations, scientists predict that the greenhouse effect will be enhanced and the Earth's climate will become warmer (Lockwood, 1979). In order to understand the present and past climate, it is important to take into consideration the various direct and indirect causes in the form of natural and anthropogenic factors :

### *(a) Sunspot Activity and Solar Flux*

The climatic fluctuations on a time scale upto thousands of years may result from variations in solar activity, generally measured as a function of 'sunspot' activity.

### *(b) Volcanic Activity*

Volcanic aerosols tend to block sunlight and contribute to short-term cooling. Volcanoes also emit carbon dioxide ($CO_2$), a greenhouse gas, which has a warming effect. It influences both long and short-term global climate. Over a long period, increased volcanic activity can emit enormous volumes of GHGs and increase the rate at which new oceanic crust is generated. Volcanic ashes could change the absorption properties of the atmosphere and this influences the global temperature over a short period (NRC, 2005).

### *(c) Continental Positions and Mountain Building Events*

Very long time changes in global climate over millions of year appear to be strongly controlled by the position of the continents. As the plates, which make up the outer surface of the earth, relentlessly move around at speeds typically measured in mm to cm per year, the size and position of the continents on the globe change.

### *(d) Anthropogenic Causes*

Human activity can lead to changes in atmospheric composition and, hence, the nature of radioactive forcing. For example, the burning of fossil fuels increases the number and distribution of aerosols, besides altering the composition of the atmosphere. Deforestation and changes in land use can also give rise to changes in climate. The amount of CO in the atmosphere has increased by more than 25% during the last century and since the beginning of the industrial revolution. This is largely a result of combustion of fossil fuels and the large scale unplanned removal of forest cover. Human activities also affect the amount of aerosols in atmosphere, which influences the climate in other ways. Man made aerosols remain in the air typically for a few days only and they tend to be concentrated near their sources such as industrial regions (IPPC, 1996).

Greenhouse gas emissions from industry, transportation and agriculture have played a major role in the recently observed global warming (EPA's Clean Air Markets-Climatic Change). Carbon dioxide, methane, nitrous oxide and three groups of fluorinated gases (but not CFCs) are the subject of the Kyoto Protocol, which entered into force in 2005. CFCs are controlled by the Montreal Protocol, motivated by their effect on stratospheric ozone, rather than by their effect on greenhouse warming.

### *(e) The Role of Water Vapour*

Water vapour is a naturally occurring greenhouse gas and accounts for the largest percentage of the greenhouse effect. Water vapour concentrations fluctuate regionally, but human activity does not directly affect water vapour concentrations except at very local scales.

In climate models, an increase in atmospheric temperature caused by the greenhouse effect due to anthropogenic gases will, in turn, lead to an increase in the water vapour content of the troposphere, with approximately constant relative humidity. The increased water vapour, in turn leads to an increase in greenhouse effect and, thus, further increase in temperature; the increase in temperature leads to still further increase in atmospheric water vapour; and the feedback cycle continues until equilibrium is reached. Thus, water vapour acts as a positive feedback to the forcing provided by human-released greenhouse gases such as $CO_2$ (Kiehl and Kevin, 1997).

### *(f) Contribution of Green House Gases Emissions from Land Clearing*

Forests are among the most carbon-dense ecosystems, comprising approximately 75% of live organic matter. When soils are included, forests hold almost half of the carbon of the world's terrestrial ecosystem

(Houghton and Skole, 1993). Although some changes in land use increase the amount of carbon stored, the trend over the past 300 years has been to decrease the area in forests. Woods Hole Research Center estimates that annually 1.6 Gt of carbon are released from changes in land use, resulting in a net decline in terrestrial carbon stocks. Since 1850, the total carbon emissions from land use change are of the order of 136 Gt.

Deforestation can take a number of forms, each of which has different implications for carbon emissions. If the forest is burned, this results in a direct emission of greenhouse gases into the atmosphere. Remote sensing images show dramatic increase of carbon monoxide emissions during the dry season over the Brazilian Amazon, when farmers are preparing their fields and ranchers are clearing forest for cattle (Shukla *et al.*, 1990).

Other forms of deforestation may have a less significant impact on climate change. For instance, deforestation for firewood collection may result in a net carbon balance, because as trees are cut, new trees take their place and begin absorbing $CO_2$ through photosynthesis. Deforestation for construction materials or furniture can have varied effects, depending on how long the wood is "locked up" in some useable form. Plywood for concrete molds may be discarded soon after it is used, thereby immediately contributing to emissions through burning or decay, whereas wood locked up in furniture or building construction may remain that way for years. According to the Woods Hole Research Center, from 1850 to 1990 about 107 Gt of carbon were lost from terrestrial ecosystem, but 10 Gt accumulated in wood products (Myers, 1988).

Conversion of natural ecosystems to croplands and pastures has resulted in net releases of 73 Gt. These agro-ecosystems continue to take up carbon, but at levels generally inferior to the previously forested ecosystems because their carbon density is far lower. Furthermore, harvested crops and decaying residues in the fields eventually result in releases of carbon back into the atmosphere.

For its heat trapping potential and its significance in global biogeochemical cycles, carbon dioxide is one of the most important greenhouse gas emissions as well. Aber and Mellilo (1991) list 10 additional greenhouse gases that are released by terrestrial ecosystems, many of which have much greater heat-trapping capacity than carbon. Among others, these include:

- methane ($CH_4$), which is emitted by wetlands, rice fields and animals.
- nitrous oxide ($N_2O$) and nitrogen oxides ($NO_x$) which is emitted by fertilized farming and biomass burning.
- ammonia ($NH_3$), which is emitted by animals, fertilized agriculture and biomass burning.
- carbon monoxide (CO) from biomass burning.
- sulphur gases from wetlands, wet tropical forests, oceans, fertilized agriculture, biomass burning.
- water vapour from forests.

Those with longest atmospheric lifetimes have received greater attention because of the potential for increased production to result in lasting increases in atmospheric concentrations. After carbon dioxide, methane and nitrous oxides are thought to have the greatest potential to trap outgoing long-wave radiation, thus contributing to the greenhouse effect and, ultimately, leading to climatic change.

***(g) Related effects***

Carbon monoxide has an indirect radiative forcing effect by elevating concentrations of methane and troposphere *ozone* through chemical reactions with other atmospheric constituents (e.g., the hydroxyl radical, OH) that would otherwise destroy them. Carbon monoxide is created when carbon containing fuels are burned incompletely. Through natural processes in the atmosphere, it is eventually oxidized to carbon dioxide. Carbon monoxide has an atmospheric lifetime of only a few months and, as a consequence, is spatially more variable than longer-lived gases.

Another potentially important indirect effect comes from methane, which in addition to its direct radiative impact also contributes to ozone formation. Shindell *et al.* (2005) argue that the contribution to climate change from methane is at least double the previous estimates as a result of this effect (Smith *et al.*, 1997).

## IMPACT OF CLIMATIC CHANGE

In fact, humans are increasingly being recognized as a dominant force in global environmental change (Vitousek *et al.*, 1997; Moran, 2001; Turner, 2001; Lambin *et al.*, 2001). Climate change and variability in the past are seen in the geological climate records. This indicates large fluctuations in global and regional climates over the past million years.

Short-term fluctuations over land areas vary widely, but broad trends on a larger scale show the globe has warmed by about 0.5°C over the past 100 years (WMO, 1994).

The impacts are mainly felt through the changes in temperature, wind patterns, precipitation and humidity. Indirect effects are mainly the consequences of sea level rise resulting from the melting of glaciers and thermal expansion of sea water. The sea level rise will further lead to erosion, flooding, salinization and other general hazards. As a result, there will be an increase in the occurrence of extreme events in the form of storms, droughts, forest fire and erosion. Some of these long-term impacts are discussed below :

**1. Land Cover Change and Loss of Productivity**

Land use is the term that is used to describe human uses of the land. It includes such broad categories as human settlements, protected areas and agriculture. Within those broad categories are more refined categories, such as urban and rural settlements, irrigated and rainfed fields, national parks and forest reserves and transportation and other infrastructure. Land cover refers to the natural vegetative cover types that characterize a particular area. Examples of broad land cover categories include forest, tundra, savannah, desert or steppe, which in turn, can be sub-divided into more refined categories representing specific plant communities (e.g., oak-pine scrublands, mangroves, seasonally flooded grassland, etc.)

The scientific research community called for substantive study of land-use and land-cover changes during the 1972 Stockholm Conference on the Human Environment, and again 20 years later at the 1992 United Nations Conference on Environment and Development (UNCED). In the past decade, a major international initiative to study land-use and land-cover change (LUCC), the LUCC Project, has gained great momentum in its efforts to understand driving forces of land-use change (mainly through comparative case studies), develop diagnostic models of land-cover change, and produce regionally and globally integrated models (Lambin *et al.*, 1999 and Geist, 2002).

The location of today's terrestrial ecosystems is the result of multiple factors, including latitude, altitude, topography, geology, soils and climate. If projected changes in temperature and rainfall occur, there will be a displacement of ecosystems from their current locations, in some cases by hundreds of kilometers. This has left some conservationists wondering if the current network of protected areas will adequately protect species that exhibit limited ranges (so-called endemics), or if species even will be able to migrate fast enough to keep up with climatic change. The climate change has an adverse impact on biodiversity and also leads to desertification.

Agriculture currently accounts for 24 per cent of the world output, employs 22 per cent of the global population, and occupies 40 per cent of the landmass. Decrease in agricultural production is often mentioned as potentially the most worrisome consequence of climatic change. This is because the components of natural ecosystem are very much sensitive to changes in weather and climate, particularly to extreme weather events, decreased soil moisture, temperature change and increase in carbon dioxide in the atmosphere. So, vegetation as well as agriculture are likely to be affected from such changes in weather and atmosphere. Due to climatic change, coastal cropland in countries, such as Bangladesh and Egypt has been extremely vulnerable to storm surges. Such events can become more common and devastating because global warming will cause sea levels to rise and may intensify storms. The green house effect will also change precipitation patterns and soil moisture, while this may benefit some agricultural regions, others will suffer. Many plants grow faster and larger in a warm environment rich in $CO_2$, and they often use water more efficiently. But optimistic estimates of greatly increased crop yields have usually been based on laboratory experiments that have ignored the influence on yields of more frequent extreme climate events (especially drought and heat waves), increased pest infestation and the decreased nutritional quality of crops grown in carbon dioxide enriched atmosphere.

In tropical regions, even small amounts of warming will lead to decline in yield. In higher latitudes, crop yields may increase initially for moderate increase in temperature but then fall. Higher temperatures will lead to substantial declines in cereal production around the world, particularly if the carbon fertilization effect is smaller than previously thought, as some recent studies suggest. Food production will be particularly sensitive to climate change, because crop yields depend, in large part, on prevailing climatic conditions, i.e., temperature and rainfall patterns. Low levels of warming in mid to high latitudes (US, Europe, Australia, Liberia and some parts of China) may improve the conditions for crop growth by extending the growing season and opening up new areas for agriculture. Further, warming will have increasingly negative impacts - as temperature thresholds are reached more often and water shortages limit growth in regions such as southern Europe and western USA. High temperature episodes can reduce yields by upto half if they coincide with a critical phase in the crop cycle such as flowering (Ramankutty and Foley, 1999).

Aside from direct impacts of climate change on crop yields (Rosenzweig and Iglesias, 2000), climate change may also result in migration of agro-ecosystems in ways that can effect global food production (Walker and Steffe, 1997). The US National Assessment on climate change impacts (USGCRP, 2000a, 2000b) devotes substantial space to the question of climate change and agriculture, and finds that the southwest's forests and agricultural lands may be transformed into a savanna and grassland due to warming and drying, and that the currently arid areas of the southwest are likely to become moisture. Fischer *et al.* (2001) predict migration of agro-ecosystems northwards in the northern hemisphere, which will benefit some temperate countries, particularly those with large frontier leads to the north (e.g., Russia and Canada).

According to the IPCC, during the last 100 years, sea level has risen at a rate of 1-2 $mmyr^{-1}$, and is projected to rise 9-88 cm by the end of the present century. Sea level rise resulting from melting glaciers, may result in a net loss of land as oceans inundate low-lying coastal areas. Coastal flooding from sea level rise is already occurring in many regions that are affected by land subsidence (e.g., northeastern USA, the Mississippi delta and eastern China). Some island states, such as Kiribati, Tuvalu, the Marshall islands and the Maldives, have their highest elevations only 4 metres above mean sea-level and are, therefore, highly vulnerable to projected sea-level rise. A rise would pose grave problems for small Island Developing States (SIDS). A large part of the Majuro attol, for example, which is home to 50 per cent of the population of the Marshall Islands, and significant landmass in Kiribati would be lost. Some cays of the Bahamas, and the Turks and Caicos islands in the Caribbean might be inundated. Sea-level rise would lead to rapid land loss through submergence of lowlands, loss of wetlands and erosion of coastal regions and beaches, alterations in delta and estuarine environments and farmlands loss in most SIDS (Eisma, 1995; Titus and Richman, 2001). A number of studies have sought to estimate the number of people vulnerable to sea-level rise (Gornitz, 2002; Small and Cohen, 1999). Other studies have focused on the vulnerability of coastal cities to sea level rise (Schiller *et al.*, 2001 and Nicholls, 1995). Due to trade and relations with agricultural hinterlands, many large urban agglomerates have developed on coastlines that may be endangered.

**2. Growth and Development of Society**

Equally important is the impact of these regional and global changes on society. The impact will not be felt evenly across the globe. Although some parts of the world would be benefitted from modest rises in temperature, but if temperature increases at high rate, most countries will suffer heavily and global growth will be affected adversely. For some of the poorest countries, there is a real risk of being pushed into a downwards spiral of increasing vulnerability and poverty. Average global temperature increase of only 1-2°C would lead to 15-40 per cent of species to extinction. Take, for example, conversion of forested areas to crop lands, pasture or human settlements. Deforestation can result in the loss of biodiversity, especially in the tropics; biodiversity loss results in declines in ecosystem integrity, and also genetic losses that may impede future scientific advances in agriculture and pharmaceutics. Deforestation can also impact hydrological processes, leading to localized declines in rainfall, and more rapid runoff of precipitation, causing flooding and soil erosion. And finally, scientists have come to a better understanding of the role - that forests play in the carbon cycle, and how forest burning in certain parts of the world is an important contributor to greenhouse gases that contribute to climate change. Clearly, all of these changes impact society. The impact on society is estimated through possible effects on water, food, health, land, environment, etc., (Sheikh and Mandar, 2008).

This dual role of humanity in, both, contributing to the causes and experiencing the effects of global change processes emphasizes the need for better understanding of the interaction between human and the terrestrial environment. This need becomes more imperative as changes in land use become more rapid. Understanding the driving forces behind land-use changes and developing models to simulate these changes are essential to predicting the effects of global environmental change (Veldamp and Lambin, 2001).

**3. Water Famine**

People will feel the impact of climatic change most strongly through changes in the distribution of water around the world and its seasonal and annual variability. Water is an essential input for all lifeforms and a requirement for good health and sanitation. It is a critical component for raising crops, sustainable growth and poverty reduction. Globally, around 70 per cent of all freshwater supply is used for irrigating crops and providing food. Climatic change will alter patterns of water availability and river flows drastically because the rise in temperature may lead to the melting of snow, resulting into drying up of the snow-fed rivers. The depletion of water in one country will have consequences in the other. For instance, the exploitation of ground water aquifers in one country may lead to depletion in the other. Also, overuse of water for irrigation in one nation can lead to its shortage in the other. Droughts and floods will become more severe in many areas. There will be more rain at high latitudes, less rain in the dry sub-tropics, and uncertain but probably substantial changes in tropical areas. Hotter land surface temperatures induce more powerful evaporation and, hence, more intense rainfall, with increased risk of flash flooding. Moreover, floods can also be caused in neighbouring countries if one of them releases excess water to control rising river levels in its territory. Differences in water availability between regions will become increasingly pronounced. Areas that are already relatively dry, such as the Mediterranean basin and parts of Southern Africa and South America, are likely to experience further decreases in water availability, e.g., several climate models predict upto 30 per cent decrease in annual run-off in these regions for a 2°C global temperature rise and 40-50 per cent for 4°C. In contrast, South Asia and parts of Northern Europe and Russia are likely to experience increases in water availability (runoff), i.e., 10-20 per cent increase for a 2°C temperature rise and slightly greater increase for 4°C, according to several climate models.

**4. Agriculture**

Agriculture's dependence on climate makes this sector an ideal one for examining potential impacts of climate change. The increase in greenhouse gases has the potential for global warming. The global and regional warming estimates differ from model to model and scenario to scenario for the increase of greenhouse gases. The impact of climatic change on agriculture is in following ways :

- Increase in the concentration of $CO_2$ would affect photosynthesis by agricultural plants, increasing the foliage and the biomass of plants and vegetation.
- Increase in temperature may cause thermal stress.

- Changes in rainfall, cloudiness (changes in solar radiation), storm activity, run-off, and extreme rainfall events may cause changes in agricultural productivity.
- Overall, due to climate change a decrease in agricultural output for India as a whole is predicted after 2050, the loss estimates varying between 13 to 25 per cent both for rice and wheat.
- Increased rainfall (even by 5%) may be beneficial to *kharif* crops but lower rainfall (even by 5% as predicted by certain models) may decrease rice yield by about the same percentage. However, higher monsoon rainfall may make *kharif* crops more vulnerable to pests and diseases, and
- Higher (lower) cloudiness associated with higher (lower) monsoon rainfall may have negative (positive) effect on photosynthesis and influence output of *kharif* crops.

**5. Impacts on Land and Coastlines**

Low-lying coastal areas are more vulnerable to the penetration of sea water in land in association with high tides, tropical storms, storm surges, etc. Wave actions caused by strong winds also lead to coastal erosion. Heavy rainfall, and resulting floods, in the coastal zones result in enhanced land erosion and the sediments are deposited near coastal belts. Accumulation of sediments over decades changes the coastline features and, in turn, result in greater penetration of sea water inland and higher inundation of coastline areas. This causes salinization of water table in low-lying areas, and even damages natural vegetation and destroys crops.

**6. Health**

Climate change will increase worldwide deaths from malnutrition and heat stress. Slum populations in urban areas are particularly exposed to disease, suffering from poor air quality and heat stress especially with limited access to clean water. It would affect the potential transmission of vector borne diseases such as malaria and dengue. Potential health effects due to increased exposure to ultra violet radiation because of the depletion of atmospheric ozone are diseases like conjunctivitis and cornea diseases of eye, cataract and skin cancer, leprosy and genetic change or mutation (NRC, 2005).

Climate change will amplify health disparities between rich and poor parts of the world. The World Health Organization (WHO) estimates that climate change since the 1970s is already responsible for over 1.5 million deaths each year through increasing incidence of diarrhoea, malaria and malnutrition, predominantly in Africa and other developing regions. Just a 1 °C increase in global temperature above pre-industrial could double annual deaths from climate change to at least 3.0 million. These figures do not account for any reductions in cold-related deaths, which could be substantial. At higher temperatures, death rates will increase sharply, for example more than a million of people are dying from malnutrition each year.

As a result of increased economic activities of different types in response to international trade and commerce, there is migration of population to coastal areas, and people have started even inhabitating low level coastal areas within the tidal zone. Rise in sea level is likely to endanger their habitat. Several National Study Team Reports have yielded the following data for land loss and population displacement for 100 cm sea level rise scenarios with no adaptation (Table 3).

**Table 3: Land loss and population displacement**

| *Country* | *Sea level rise (cm)* | *Land loss area* | | *Population displaced* | |
|---|---|---|---|---|---|
| | | *km²* | *%* | *Million* | *%* |
| Bangladesh | 100 | 29846 | 20.7 | 14.8 | 13.5 |
| India | 100 | 5763 | 0.4 | 7.1 | 0.8 |
| Pakistan | 100 | 1700 | 0.2 | - | - |
| Malaysia | 100 | 7000 | 2.1 | 0.05 | 0.03 |

It is clear from the data depicted in table 3 that Bangladesh is one of the most endangered country. Land loss due to erosion would affect several smaller islands, which are more vulnerable to sea level rise.

### 7. Environment

There are many environmental problems coming from the increase in concentration of greenhouse gases in Earth's atmosphere. The Earth's climate has changed due to increased water vapour in the atmosphere, glaciers and polar ice caps appear to be melting, floods and droughts are becoming more severe and sea levels have risen, on average, between 4 to 10" since 1990. Global warming has caused more floods and more droughts. The rises in sea levels can increase the salinity of freshwater throughout the world, and cause coastal lands to be washed under the ocean. Warmer water and increased humidity may encourage tropical cyclones, and changing wave patterns could produce more tidal waves and strong beach erosion on the coasts.

Climate change is likely to occur too rapidly for many species to adapt. A study reports that with a rise of 2°C, 15-40 per cent of species face extinction. Severe drying over the Amazon region, as predicted by some climate models, would result in dieback of forest with the highest biodiversity on the planet. The warming of the 20th Century has already directly affected ecosystems. Over the past 40 years, species have been moving pole wards by 6 km on average per decade, and seasonal events, such as flowering or egg-laying have been occurring several days earlier each decade. Coral bleaching has become increasingly prevalent since the 1980s. Arctic and mountain ecosystems are acutely vulnerable-polar bears, caribou and white spruce have all experienced recent declines. Climate change has already contributed to the extinction of over 1 per cent of the world's amphibian species from tropical mountains. Ecosystems will be highly sensitive to climate change. For many species, the rate of warming will be too rapid to withstand. Many species will have to migrate across fragmented landscapes to stay within their "climate envelope". In some cases, the "climate envelope" of a species may move beyond reach, for example, moving above the tops of mountains or beyond coastlines. Conservation reserves may find their local climates becoming less amenable to the native species. Other pressures from human activities, including land-use change, harvesting/hunting, pollution and transport of alien species around the world, have already had a dramatic effect on species and will make it even harder for species to cope with further warming. Since 1500 A.D., 245 extinctions have been recorded across most major species groups, including mammals, birds, reptiles, amphibians, and trees. Rising levels of $CO_2$ have some direct impacts on ecosystems and biodiversity, but increases in temperature and changes in rainfall will have even more profound effects. Vulnerable ecosystems are likely to disappear almost completely at even quite moderate levels of warming. The arctic will be particularly hard hit, since many of its species, including polar bears and seals, will be very sensitive to the rapid warming predicted and substantial loss of sea ice.

## Ecological Effects of Ozone Layer Depletion

Increasing amounts of UV radiation will have an impact on plankton and other tiny organisms at the base of the marine food web. Ozone layer depletion seems likely to increase the rate of greenhouse warming by reducing the effectiveness of the carbon dioxide sink in the oceans. Phytoplankton in the oceans assimilates large amounts of atmospheric carbon dioxide. Increased UV radiation will reduce phytoplankton activity significantly.

A high increase in UV radiation (280-315 nm) may disrupt many ecosystems on land. Rice production may be drastically reduced by the effects of UV-B on the nitrogen assimilating activities of micro-organisms. With a diminishing ozone layer, it is likely that the supply of natural nitrogen to ecosystems, such as tropical rice will be significantly reduced. Most plants and trees grow more slowly and become smaller and more stunted as adult plants when exposed to large amounts of UV-B. Increased UV-B inhibits pollen germination.

UV-B stimulates the formation of reactive radicals - molecules that react rapidly with other chemicals, forming new substances. The hydroxyl radicals, for example, stimulate the creation of tropospheric ozone and other harmful pollutants. Smog formation creates other oxidized organic chemicals such as formaldehydes. These molecules can also produce reactive hydrogen radicals when they absorb UV-B. In urban areas, a 10 per cent reduction of the ozone layer is likely to result in a 10-25 per cent increase in troposphere ozone. More UV-B radiation seems likely to cause global increases in atmospheric hydrogen peroxide. This is the principal chemical that oxidizes sulphur dioxide to form sulphuric acid in cloud water, making it an important part of acid rain formation (Singh *et al.*, 2007).

UV-radiation also causes many materials to degrade more rapidly. Plastic materials used outdoors will have much shorter lifetimes with small increases of UV radiation. PVC sidings, window and door frames, pipes, gutters, etc., used in buildings degrade faster.

## International Conventions Towards the Global Climatic Change

Since the atmosphere is one entity and covers the entire globe, it is impossible to handle the problem of climate change only by a single nation or a group of nations. There have been various organized efforts at all levels on a global basis to observe and study climate. In these, World Meteorological Organization, a UN specialized agency took a leading role. Some of the institutional frameworks through which global efforts are being channelised are described below (Kumar and Parmar, 2005):

**1. Intergovernmental Panel on Climatic Change (IPCC)**

The World Meteorological Organization (WMO, 1994) and the United Nations Environmental Programme (UNEP) jointly established IPCC in 1988 in order to assess the available scientific information on climatic change, estimate the environmental and socio-economic impacts of climate change and formulate response strategies. This panel has brought out many authoritative reports focusing on different aspects of climate change. According to its 1995 report on climate change, there has been increase in greenhouse gases concentration, which can be largely attributed to human activities, mainly due to use of fossil fuel, industrial activities, change in land use and agriculture.

**Agenda 21**

The United Nations Conference on Environment and Development, which took place in Rio de Janerio, Brazil in 1992, is an event of historic proportions. Agenda 21 adopted in this Conference is a 'blue print' for action in all major areas affecting the relation between the environment and economy. It focused on the period upto the year 2000 and extends into the 21st century. Some of the proposals are as follows :

- Understanding the natural environment in its totality,
- Development of early warning system,
- Prioritize areas of research,
- Governmental functionaries should identify the most vulnerable areas to erosion, floods, landslides, earthquakes, etc., and
- Partnership between environmental and developmental agencies.

**2. Global Change and Terrestrial Ecosystem (GCTE) Project**

It is core project of the International Geo-sphere Biosphere Programme (IGBP), aimed at developing capacity to predict the effects of changes in climate, atmospheric composition and land use practices on terrestrial ecosystem.

### 3. Global Climate Observing System (GCOS)

It is an observing system/programme recommended by the Scientific/ Technical Committee of the second World Climate Conference and endorsed by the Eleventh Congress of WMO, which met in Geneva in May 1991. Global Climate Observing System is intended to meet the needs for - climate system monitoring, climate change detection and monitoring of the response to climate change, especially in terrestrial ecosystem, and mean sea land data for application to national economic development research towards improved understanding, modeling and prediction of the climate system.

### 4. Global Environmental Monitoring System (GEMS)

Established in 1975, Global Environmental Monitoring System is administered by UNEP and is a monitoring system on various aspects of global environment including climate, atmosphere, oceans and coastal areas.

### 5. World Climate Programme (WCP)

Initiated under the aegis of World Meteorological Organization and UNEP, this programme promotes the use of climate information to assist economic and social planning and monitoring of climatic variations to take up both corrective and preventive measures. It has four components :

- World Climate Data Programme (WCDP) assists countries in setting up climate data monitoring system and acquiring data processing capability that could help economic policy making.
- World Climate Research Programme (WCRP), the objective of which is to improve our knowledge of climate, climate variation and the mechanisms that bring about climate change to the extent climate change can be predicted and man's influence on climate delineated. This will enable us to predict climate variations in advance so as to provide useful climate forecasts to different user interests.
- World Climate Applications Programme (WCAP) assists both developing and developed countries to collect, analyse and apply climate information to economic sector such as agriculture, water resource use, energy generation and health.
- World Climate Impact Studies Programme (WCIP) under which several projects have been undertaken and expert groups set up to study the

socio-economic impact on climate fluctuation and change. The objective of the WCIP is to improve the methodology of undertaking assessment of climate impact.

**5. World Weather Watch (WWW)**

It is a system created by WMO for collecting, processing and distributing throughout the world, weather and other environmental information. This provides a solid foundation for the World Climate Programme to function.

Several conferences in the recent years have taken place, which have provided international policy framework to be considered when dealing with the science of the global climate change.

During the Earth Summit, success in a limited form was seen in the different treaties signed by the attending nations. Two key treaties were signed by over 150 nations - the treaty on biodiversity and climate change. An agreement was signed by the countries on the statement, and so has the setting up of the high level commission on sustainable development to monitor the implementation of Agenda 21 by the countries. As for the funding of the cleaning up of environment, Japan during the Summit said, it will increase its development aid by 400 million a year to 1.4 billion, Germany and France promised to raise their contributions to 30.7% of their gross national products by the year 2000. Some 150 nations signed a treaty that might, when carried out, decrease the emissions that contribute to global warming at the 1990 level by the year 2000. In 1987, the signing of the Montreal Protocol agreement by forty-six nations established an immediate timetable for the global reduction of chlorofluorocarbons production and use.

One of the major conventions concerning global warming resulted in the Kyoto Protocol, held in Kyoto, Japan, between December 1-11, 1997. Delegates from all over the world were present in order to find a universal agreement to reduce greenhouse gas emissions. The results had most developed nations doing most of the reducing; the United States must cut emissions 7%, Japan 6% and the European Union 8% below 1990 levels.

The United States proposed a plan to have these levels cut over a five year period between 2008-2013. The United States also said it will not sign the protocol if other developing/undeveloped countries do not sign it as well, fearing the economy will falter. The US was successful in emissions trading with other countries, and can purchase emission permits from other countries who have extra permits. This stresses the importance of flexibility the US was looking for when it said it can not lower the emission levels until at least 2008. Again, the US is trying to look out for its own economy first. If a country shall fail in completing its goal, the country will then not be able to receive joint implementation projects. However, this protocol is not yet law; it must be ratified by at least 55 countries, accounting for 55% of the world's total greenhouse gas emissions.

If the Kyoto Protocol becomes the law of the land, there are potential economic problems that may lead to a change in quality of life for many Americans. By reducing greenhouse gas emissions, people will be more healthy due to better air quality and water quality. However, there may be a reduction in the rate of economic development because industries will have to adapt and find different ways of producing goods. People will have to drive smaller, lighter cars, ride bicycles more often, and increase efficiency in many ways (US Dept. of State, 1998).

The eighth Conference of Parties (CoP-8) to the United Nations Framework Convention on Climate Change (UNFCCC) was held recently in New Delhi. It is an appropriate occasion to ponder over issues such as what is climatic change and how does it affect us. Durban (2000) and several other protocols have also been organized.

## STEPS FOR COMBATING CLIMATE CHANGE

### 1. Sequestration of Carbon Through Land Management

In the past decade, there has been growing interest in mitigating human impacts on the climate through activities in two principal domains: emissions reductions through increased fuel efficiency and sequestration of carbon in terrestrial ecosystems. There is also on-going research into how to increase carbon sequestration in rocks and oceans. The IPCC's Special Report on Land-use, Land-use Change, and Forestry is dedicated to an assessment of the feasibility, from scientific, technical and institutional perspectives, of withdrawing carbon from the atmosphere through the use of terrestrial sinks (IPCC, 2000).

The largest terrestrial reservoirs of carbon are tropical and boreal forests, which store 428 and 559 Gt of carbon, respectively, in above and below ground biomass. Savannas and grasslands also store large amounts of carbon (over 300 Gt each), wetlands store 240 Gt, and temperate forests store 159 Gt (IPCC, 2000). Because of the importance of forests as carbon sinks, most of the policy attention has been focused on how to increase carbon uptake through these ecosystems.

Trees grow and gain carbon when the amount of carbon assimilated through photosynthesis exceeds the amount of carbon lost from respiring leaves, branches, stems and roots (Barnes *et al.,* 1998). At the forest ecosystem level, the net carbon uptake or emission will be determined by the balance of plant photosynthesis and respiration, as well as respiration by decomposers (Aber and Millilo, 1991). Newly planted or regenerating forests will continue to uptake carbon for 20 to 50 years after establishment, depending on species and site conditions. The rates vary, depending on forest type and latitude, from a low of 0.4-1.2 tonne $ha^{-1}$ in boreal regions to a high of 4-8 tonne $ha^{-1}$ $yr^{-1}$ in tropical regions. A forest ecosystem, as a whole, will tend toward a zero carbon balance at maturity.

In July 2001, the contracting parties of the UNFCCC, with the exception of the United States, reached agreement on a plan for implementing the Kyoto Protocol. The Protocol obliges industrialized countries to reduce their greenhouse gas emissions, and create, innovative mechanisms by which emissions allocations can be traded among states. In addition, parties may meet their emissions targets by sequestering carbon in forests. The Protocol's Clean Development Mechanism permits parties to invest in afforestation and reforestation projects in developing countries. A number of developing countries have been exploring the possibility of entering the carbon market by protecting tropical forests or reforesting large areas (de Sherbinin *et al.,* 2002).

One such effort is taking place in the Noel Kempff Mercado National Park in northeastern Bolivia. The Park covers more than 3.7 million acres of dense tropical rainforest, and the project is expected to sequester 7 million to 10 million tonnes of carbon during its 30 year life, largely through protection of existing forests and accumulation of additional biomass in currently unforested areas. As part of the justification, project proponents need to prove that the encompassed area would, otherwise, have been deforested. In other words, by protecting the forest, the project is yielding carbon sequestration benefits that would not otherwise have accrued, had the area been left open for logging. Similar pilot projects are testing approaches to carbon sequestration in Costa Rica, Uganda and several industrialized nations.

Although the potential for carbon sequestration in forests is quite significant, the way forward is fraught with difficulties (IPCC, 2000). The challenges have to do with accurately measuring carbon uptake that would not have occurred, apart from the project intervention or leakage of emissions that occurs when one forest is adequately protected but the forest cutting activity is simply displaced to another area that is not subject to such restrictions. There are also technical matters of setting up a carbon accounting system, and identifying appropriate tools (such as remote sensing) so that carbon monitoring over vast tracts of land can be done in a cost effective manner. Some have also voiced concern that monocropped tree plantations might supplant natural forests, with their biodiversity and ecosystem values. Finally, there is the issue of permanence. Though sequestration may be useful in the short-term , for slowing the rate of climate change by soaking up 'surplus' carbon dioxide, protecting a forest indefinitely would require institutional arrangements that would also last indefinitely, something that humans have never attempted.

Globally, drylands are more important than temperate forests in terms of carbon storage (199 Gt versus 159 Gt). Some experts see great promise for linking actions to combat desertification to actions that would mitigate against climate change through carbon sequestration (Olson *et al.*, 2001). Dryland agricultural soils are extensively degraded in many parts of the world, which means they lack nutrients and organic matter. Increasing nutrients and organic matter is good not only for increasing crop productivity; it also happens to lead to increases in soil carbon. By judicious use of chemical fertilizers and organic matter inputs, some see the possibility for reversing the steady decline in agricultural productivity in some regions, while simultaneously stockpiling carbon in agricultural soils (USGS, 2000 and Woomer *et al.*, 2001).

There are many other land management activities that can enhance carbon uptake and a number of other land-use change activities. These are enumerated in table 4. Such activities are being considered under Article 3.4 of the Kyoto Protocol.

**Table 4: Potential "Additional Activities" under Article 3.4 of the Kyoto Protocol**

| *Land-use* | *Activity* |
|---|---|
| ***A. Improved management within a land-use*** | |
| Cropland | Reduced tillage, rotations and cover crops, fertility management, erosion control, and irrigation management |
| Rice | Irrigation, chemical and organic fertilizer, and plant residue management |
| Agroforestry | Better management of trees and cropland |
| Grazing land | Hard, woody plant, and fire management |
| Forest land | Forest regeneration, fertilization, choice of species, reduced forest degradation |
| Urban land | Tree planting, waste management, wood product management |
| ***B. Land-use change*** | |
| Agroforestry | Conversion from unproductive cropland and grasslands |
| Restoring severely | To crop, grass or forest land degraded land |
| Grassland | Conversion of cropland to grassland |
| Wetland restoration | Conversion of drained land back to wetland |

***Source:*** IPCC, 2000

## 2. Coastal Zone Management

Coastal zone is the area of interaction between the land and the sea. Both terrestrial as well as marine environments influence this zone. Interactions between various natural processes and human activities are important factors in the coastal areas. Population pressure has resulted in the increased exploitation of coastal resources. Also, the increase in the load on ports due to expanding international trade, location of waste disposal sites, development of chemical, petrochemical, fertilizer and allied industries and brackish water aquaculture have resulted in further exploitation of coastal resources.

The major issues involved in coastal management are coastal habitat, coastal processes, and water quality. Coastal habitats consist of wetlands, beaches, lagoons, marshes, coral reefs, mangroves, sand dunes, etc. Coastal processes like erosion, deposition, sedimentation transport, flooding and sea level changes are continuously at work to modify the shore line. Water quality is the lifeline of coastal habitat. Several bays and estuaries are being used to dump waste materials into the sea and these activities inadvertently degrade water quality. There are other upstream activities such as agriculture, small dams for power and irrigation, urbanization etc., which also degrade water quality. Excess nutrients over fertilize marine environments, causing algae blooms (including toxic algae) that cause damage to fishery resources (Gornitz, 2002).

Threat of sea level rise due to climate change has focused attention on coastal zones and small islands and created awareness about the vulnerability of these regions, particularly to human population and other ecosystems habitating the low lying coasts, tidal deltas, small islands, atolls, etc. Three groups of response strategies can be distinguished, i.e. (i) planned retreat of population and restriction of development activities in vulnerable coastal areas/islands (ii) adaptive response, such as elevation of buildings, modification of drainage systems and land use changes to allow for the possible increase in landless due to sea level rise, and (iii) construction of defensive structures and strengthening of existing structures, if any to maintain the shorelines at their present position. To adopt these response strategies, individual countries have to undertake appropriate surveys for suggesting policy initiatives to implement the appropriate response strategy alternatives, in the context of coastal zone management plans and constraints of financial resources. A variety of impacts are involved in the coastal zones, and each specific area and sector may require different options to reduce, prevent and mitigate the likely adverse impacts associated with sea level rise. Effective adaptation to sea level rise would require a flexible coastal management strategy at different time scales that takes care of short-term and long-term goals. Assessing the full cost of adaptation to sea level rise is quite complicated and a whole range of techniques are required. Remote sensing by satellites offers an effective technique for coastal zone management to prepare databases on the following lines :

- Installation of tide gauges along vulnerable coastal areas,
- Mapping of the sea level topography by satellite altimetry near the coast,

- Detailed mapping of coastal habitats, such as wetlands, mangroves, marshes, islands, lagoons, coral reefs, beaches, etc., to document the extent of present habitats on regional and local scales, and monitor changes on year to year basis. Creation of baseline inventories of the present extent of the different habitats is a pre-requisite to changes in future with sea level rise,
- Identification of plant communities and brackish water aquaculture ponds in the coastal zones, and their periodic monitoring for determining any changes,
- Mapping of shore-line changes, such as coastal lakes, deltaic changes, erosion and sedimentation,
- Mapping of land form and land use patterns, and
- Mapping of changes in coastal salinity.

Remote sensing data with multi-spectral sensing techniques, alongwith other collateral data, can be used to monitor and predict the impact of projected sea level rise, as well as to help in designing appropriate strategies to reduce and mitigate the adverse impacts on the coastal regions. The results of remote sensing studies are to be integrated with programmes and plans for economic development, environmental quality management and land use in the coastal zones. Reliable databases are urgently needed to derive policy options in tackling the threat of sea level rise to the fragile and overexploited environment of Coastal Zones. Several international research programmes like Land Ocean Interactions in the Coastal Zones (LOCZ) under the International Geosphere-Biosphere Programme (IGBP) and the Global Sea Level Observing System (GLOSS), Global Climate Observing System (GCOS) and Global Ocean Observing System (GOOS) have been launched, and will also help in tackling the threat of sea level rise projected under the climate change scenarios.

**Some Suggestions**

- Population growth should be stabilized.
- There is a need to give technological and economic support to the poor countries.
- Provide subsidy to promote eco-friendly technologies.
- Use HCFC-134 and HCFC-134A instead of HCFC-22 and HCFC blends.
- Use CNG and electric batteries instead of diesel and petrol vehicles.
- Use electric generators/inverters instead of petrol/diesel generators.

- Use LPG cylinder instead of burning fossil fuel.
- Use electric heater's/*chulha* instead of LPG cylinder/fossil fuel.
- Reduce the use of CFC and other solvents in refrigeration.
- There is a need for developing public transport system, i.e., rail, bus instead of private vehicles, i.e., scooters, motorcycles, cars, etc.
- Subsidization of the new technologies, and allocation of funds for the research and development in this field.
- Take up massive afforestation programmes such as agroforestry and social forestry, start programmes like *Chipko, Appiko* and *Khejri* movements again to protect trees.
- Use the eco-friendly materials in the industries.
- Installing the less polluting devices in the industries, i.e., electronic precipitator for air pollution or use some other kind of technology, which reduces the pollution.
- Use of mud-soil on the top of the roof to reduce the warming instead of cemented/concrete roofs, and use of the natural chicks to protect the building from direct heat of the sun instead of glass.

## CONCLUSION

Certain atmospheric gases, such as $CO_2$, $CH_4$ and water vapour, absorb the terrestrial radiations from the heated up earth surface, leading to a rise in the temperature of the earth's atmosphere. This natural green house effect is getting exacerbated due to excessive anthropogenic emissions of such greenhouse gases. Many persistent man-made compounds, such as CFCs, interact with upper atmospheric layer of ozone gas, and deplete it through its dissociation. This ozone depletion is another cause of global warming, because the high energy, harmful UV waves of the sun are now reaching earth surface with lesser constraint, reinforcing global warming and climatic change. Climatic change is linked to emissions, and, in turn, to economic growth. Limiting emissions is then about limiting growth. Thus, sharing growth between nations has been the main bugbear.

Climate change due to greenhouse gases and ozone depletion adds to, and actively interacts with, a variety of other environmental problems by exacerbating other stressors (such as population increase or aquifer depletion), and creating new ones (such as sea level rise). Development choices in both industrialized and developing nations must consider both sides of the climate-human interaction : aiming for a moderation in greenhouse gas emissions and anticipating the impacts of expected climate trends. The crisis of climatic change is urgent and needs drastic actions to cut emissions. Community awareness, public participation and implementation of new technologies are must for environmental sustainability and preserving life on this planet.

## REFERENCES

Aber, J.D. and Melillo, J.M. (1991). *Terrestrial Ecosystems*. Sanders College Publishing, Philadelphia.

Barnes, B.V., Zak, D.R., Denton, S.H. and Spurr, S.H. (1998). *Forest Ecology*, 4th edn., John Wiley & Sons, New York.

de Sherbinin, A., Kline, K. and Raustiala, K. (2002). Remote sensing data: Valuable support for environmental treaties. *Environment*, 44(1): 50-61.

Eisma, D. (Ed.) (1995). *Climate Change Impact on Coastal Habitation.* Lewis Publishers, Springer Verlag : New York.

Encyclopedia Britannica (1988). Greenhouse Effect, Volume 5, pp. 470.

EPA (1989). The Potential Effects of Global Climate Change on the United States. Report to Congress. USEP Agency, Washington, DC. EPA 230-05-89-052.

EPA's Clean Air Markets-Climate Change (http://www.epa.gov/airmarkets/climatchg/index.html).

Falkowski, P. (2000). Integrated understanding of the global carbon cycle - A test of our knowledge. *Science,* 290 : 291-296.

Fischer, G., Shah, M., van Velthuizen,H. and Nachtergaele, F.O. (2001). Global Agro-ecological Assessment for Agriculture in 21st Century, IIASA, Luxemburg, Austria.

"Fluorocarbons" World Book Encyclopedia, Vol. 13, March, 1981, pp. 358.

Geist, H.J. (2002). The IGBP-IHDP Joint Core Project on Land-use and Land-cover change (LUCC). *In : The Encyclopedia of Life Support Systems,* Global Sustainable Development - Land-use and Land Cover. Vol. 5 (Eds. A. Badran *et al.*). UNESCO-EOLSS Publishers, Oxford, UK.

"Global Warming" United States Environmental Protection Agency (http:www.epa.gov/globalwarming/) March 8, 1998.

Gornitz, V. (2002). Coastal populations, topography and sea level rise. Goddard institute for Space Studies (http://www.giss.nasa.gov/research/ intro/gornitz 04/).

Houghton, R.A. and Skole, D.L. (1993), Carbon. *In: The Earth As Transformed by Human Action* (Eds TB.L. Turner II, W.C. Clark, R.W. Kates, J.F. Richards, J.T. Mathews and W.B. Meyer). Cambridge University Press, New York, pp. 393-408.

IPCC (2000). *Land-use, Land-use Change and Forestry.* Cambridge University Press, Cambridge, UK.

IPCC (2001) *Ecosystem and Biodiversity,* IPCC Working Group II. Third Assessment Report of the Intergovernmental Panel on Climate Change. Cambridge University Press, New York, pp. 1032.

IPCC (1990) Intergovernmental panel on climate changes. *In : Climate Change.* The IPPC Scientific Assessment, WMO/UNEP, Cambridge University Press, Cambridge, UK.

IPCC (1994). Intergovernmental panel on climate changes. *In : Climate Change.* Working group I and II of the IPPC. Cambridge University Press, Cambridge, UK.

IPCC (1996). Intergovernmental panel on climate changes, *In: Climate Change.* The second IPPC Scientific Assessment, WMO/UNEP, Cambridge University Press, Cambridge, UK.

Khalil, M.A.K., Ramussen, R.A. and Moraes, F. (1993). Atmospheric methane at Cape Mears: Analysis of a high resolution database and its environmental implications. *J. Geophysics Research,* 98: 14753.

Khan, M.Y., Biswas, M.J.C., Haque, N. and Girdhar, N. (1996). Methane emission from Indian ruminants. Proc. *National Symposium on "Prospect of Livestock and Poultry Development in 21st Century"* held at IVRI, Izatnagar, India, pp. 41.

Khan, M.Y., Khan, F. and Haque, N. (2001 ). Global warming and stratospheric ozone layer depletion by greenhouse gases with special reference to methane production from Indian livestock. *Animal Nutrition and Feed Technology,* 1(2) : 79-96.

Kiehl, J.T. and Kevin E. Trenberth (1997). Earth's annual global mean energy budget. *Bull. Am. Meteorological Soc.,* 78(2) : 197-208.

Kumar, S. and Parmar, D. (2005). Conventions and scientific programmes towards the global climatic change. *In : Environment and Development* (Ed. Jagbir Singh). I.K. International Publishing House, New Delhi, pp. 333-342.

Lambin, E.F., Baulies, X., Bockstael, N., Fischer, G, Krug Leemans, T.R., Moran, E.F., Rindfuss, R.R., Sato, Y, Skole, D., Turner, B.L. II., Vogel, C. (1999). Land use and land cover change (LUCC) : Implementation Strategy (IGBP Report, IHDP Report 10). IGBP and IHDP Secretariats, Stockholm, Bonn, pp. 125.

Lambin, E.F., Turner II, B.L., Geist, H.J., Agbola, S.B., Angelsen, A., Bruce, J.W., Coomes, O., Dirzo, R., Fischer, G, Folke, C., George, P.S., Homewood, K., Imbemon, J., Leemans, R., Li, X., Moran, E.F., Mortimore, M., Ramakrishnan, P.S., Richards, J.F., Skanes, H., Steffen, W., Stone, G, Svedin, U., Veldkamp, T.A., Vogel, C. and Xu, J. (2001). The causes of land-use and land cover change : Moving beyond the myths. *Global Environmental Change : Human and Policy Dimensions*, 11(4): 513.

Lockwood, J.G. (1979). *Causes of Climate.* John Willey & Sons, New York.

Masters, G.M. (1994). *Introduction to Environmental Engineering and Science.* Prentice Hall of India Pvt. Ltd., New Delhi, pp. 375.

"Methane" World Book Encyclopedia, Vol. 7, Friend, 1982, pp. 270.

Migeotte, M.J. (1948). Spectroscopic evidence of methane in the Earth's atmosphere. *Physical Reviews*, 73 : 519-520.

Moran, E.F. (2001). Progress in the last ten years in the study of land use/ cover change and the outlook for the next decade, *In : Human Dimensions of Global Change* (Eds. A. Dickman *et al.).* MIT Press, Cambridge, MA.

Moss, A.R. (1994} Methane production by ruminants - Literature Review of I. Dietary Manipulation to Reduce Methane Production. II Laboratory Procedures for Estimating Methane Potential of Diets. *Nutritional Abstract and Reviews (B)*, 64 : 785-806.

Myers, N. (1988). Tropical deforestation and climatic change. *Environmental Conservation*, 15(4): 293-298.

Nicholls, R. (1995). Coastal megacities and climate change. *Geo. Journal*, 37(3):369-579.

NRC (2005). *Radiative Forcing of Climate Change.* National Academy Press, Washington, DC.

Ramakutty, N. and Foley, J.A. (1999). Estimating historical changes in global land cover : Croplands from 1700 to 1992. *Global Biogeochemical Cycles*, 13(4): 917-1027.

Rosenzweig, C. and Iglesias, A. (2000). Potential impacts of climate chan on world food supply : Data sets from a major crop modeling stu CIESIN, Palisades, NY.

Schiller, A., de Sherbinin, A., Hsieh, W., Pulsipher, A. (2001). vulnerability of global cities to climate hazards. Paper presente the open Meeting of the Human Dimensions of Global Environme Change Research Community, 4-5 October, 2001, Rio de Janeir

Schimel, D., Enting, I.G., Heimann, M., Wigley, T.M.L., Raynaud, D., Alves, D. and Siegenhaler, V. (1995). $CO_2$ and the carbon cycle. *In : Climate Change.* Cambridge University Press, Cambridge, U.K.

Sheikh, M.M., Kumar, R. and Mandar, B. (2008). Climate change and society : A global perspective, Chapter 9, *In : Natural Hazards and Disaster Management* (Ed. B.C. Jat). M.D. Publication Pvt. Ltd., New Delhi, pp. 125-136.

Shindell, Drew T., Faluvegi, G; Bell, N., Schmidt, Gavin A. (2005). An emission-based view of climate forcing by methane and tropospheric ozone. *Geop. Res. Lett.*, 32(4) : 1-5.

Shukla, J., Nobre, C. and Sellers, P. (1990). Amazon deforestation and climate change. *Science*, 247 : 1322-132.

Singh, A.K., Tomar, Alka and Singh, K.K. (2007). Ozone layer depletion and its effect on environment and human health, Chapter 11, *In : Environment and Sustainable Development,* Vol. 1 (Ed. K.A. Rasure), Serials Publications, New Delhi, pp. 179-200.

Singh, A.K., Tomar, Alka, Singh, K.K. and Phogat,V. (2007). Acid rains and its harmful effects. Chapter 2, *In : Environmental Degradation and Protection,* Vol. 1 (Eds. K.K. Singh *et al.*). M.D. Publications Pvt. Ltd., New Delhi, pp. 38-78.

Singh, K.K. (2008). *Global Warming in 21st Century : Causes. Effects and Future.* M.D. Publications Pvt. Ltd., New Delhi, pp. 1-293.

Small, C. and Cohen, J. (1999). Continental physiography, climate and the global distribution of human population. *Proc. Int. Symp. Digital Earth* (http://www.ldeo.Columbia.edu/~small/population.html).

Smith, H.J., Wahlen, M. and Mastroianni, D. (1997). The $CO_2$ concentration of air trapped in GISP2 ice from the last glacial maximum-Holocene transition. *Geop. Res. Lett.*, 24(1): 1-4.

Tandon, M., Pandey, H. and Singh, V. (2005). Methane production, global warming and ozone layer depletion : A challenge for eco-friendly livestock farming, *In : Environment and Sustainable Development* (Ed. Jagbir Singh), I.K. International Publishing House, New Delhi, pp. 321-331.

Taylor, K. (1999). Rapid Climate Change. *American Scientist.* 87 : 320-327.

Titus, J.G. and Richman, C. (2001). Maps of lands vulnerable to sea level rise : Modeled elevations along the US Atlantic and Gulf coasts. *Climate Research*, 18 : 205-228.

Turner, B.L. (2001). Land-use and Land Cover Change: Advances in 1.5 Decades of Sustained International Research on Land use and Land

Cover Change, *In: Challenges of a Changing Earth* (Eds. W. Steffan *et al.*), Heidelberg, Springerverlag, pp. 21-26.

Tyler, S.C. (1991). The global methane budget. *In : Microbial Production and Consumption of Greenhouse Gases : Methane, Nitrogen Oxides and Halomethanes* (Eds. E.J. Rogers and W.B. Whitman). American Society for Microbiology, Washington D.C., pp. 7-58.

US Department of State (1998). The Kyoto Protocol on Climate Change (http://www.state.gov/www/global/oes/fs_Kyoto_climate_980115.html).

US Global Change Research Program (USGCRP) (2000a). *Climate Change Impacts on the United States.* Cambridge University Press, New York.

US Global Change Research Program (USGCRP) (2000b). Climate *Change Impacts on the United States.* Cambridge University Press, New York.

Veldkamp, A. and Lambin, E.F. (2001). Editorial : Predicting Land Use Change. Agriculture, *Ecosystems and Environment,* 85(1-3) : 1-6.

Vitousek, P.M., Mooney, H.A., Lubchenco, J. and Melillo, J.M. (1997). Human domination of earth's ecosystems. *Science,* 277(25) : 494-499.

Walker, B. and Steffen, W. (1997). An overview of the implications of global change for natural and managed terrestrial ecosystems. *Conservation Ecology,* 1(2). (http://www.ecologyandsociety. org/vol l/iss2/art2/index.html).

White, R. and McGovern, D. (1993). *Global Warming.* Prentice Hall International Ltd., London, UK.

WMO (1994). Observing the world's environment : Weather, climate and water. WMO-No. 796, World Meteorological Organization, Geneva.

Woomer, P.L., Tieszen, L.T., Tschakert, P., Parton, W.J., Toure, A. (2001). Landscape carbon sampling and biogeochemical modeling : A two-week skills development workshop conducted in Senegal. SACRED Africa, Nairobi, Kenya.

World Resource Institute (2000). Forest and Land-use Change. Carbon Sequestration Projects. WRI, Washington, DC.

# Chapter 10

# An Approach for Estimating Green House Gas Emission Inventory and Modeling in a Petroleum Refinery

*Padma S. Rao*[1], *V.M. Mhaisalkar*[2] *and S, Devotta*[3]

## INTRODUCTION

Petroleum refineries are large industrial installations that are responsible for the emission of several pollutants into the atmosphere. For most refineries, the number of products is limited and fairly well defined, however, the volumes may be different. There are refineries that make speciality products such as lubricants and solvents. On macro-scale, the crude oils vary only to a certain extent in their composition. Consequently, refinery emissions are mainly related to the raising of energy for the different processes, and type of emissions are well defined, viz., CO, $SO_2$, TSP, NOx and HC. Appropriate field data are required in evaluating the exposure of population in urban and roadside environments (Kim *et al.*, 2002). In a petroleum refinery, fugitive emissions of volatile organic compounds (VOC) are mainly from the process leakages, and off-site areas such as tankages and oily water sewer/effluent treatment system. If not well controlled. These emissions can cause a significant loss to the

[1]Air Pollution Control Division, National Environmental Engineering Research Institute (NEERI). Nehru Marg. Nagpur - 440 020, India, E-mail: ps_rao@neeri.res.in

[2]Environmental Engineering, Department of Civil Engineering, VNIT, Nagpur, India.

[3]National Environmental Engineering Research Institute (NEERI), Nehru Marg, Nagpur - 440 020, India.

refinery. In addition, they also have environmental impact, owing to the fact that they play major role in the formation of photochemical smog and ground-level ozone pollution. In resource conservation and environmental protection, there is a need to carry out inventory studies and air quality modeling of pollutants, followed by control methods, to be translated into a waste reduction/environmental protection action plan.

The major refinery processes are heating hydrocarbons for processing, physical separation and purification, chemical conversion such as residue upgrading, cooling of the products, storage of crude oil and products (Rao *et al.,* 2005). Air pollutants from refineries mainly originate from process furnaces, boilers, gas turbines, Fluidized Catalytic Cracking (FCC) regenerators, flare systems, incinerators, Sulphur Recovery Units (SRU), coke/plants, storage and handling facilities, oil/water separation systems, fugitive emissions from flanges, etc., and vents (Rao *et al.,* 2005).

Awareness of air pollution has led to numerous studies on ambient air pollutants (CPCB, 2000). Air pollutants adversely influence many atmospheric processes, including cloud formation, visibility, solar radiation, and precipitation, and play a major role in acidification of clouds, rain and fog (Hong *et al.,* 2002; Fang *et al.,* 2002). Sulphur dioxide gas contributes to the deterioration of air quality, vegetation, animal life and monuments, etc.. Several epidemiological studies have demonstrated a direct association between atmospheric inhalable pollutants and respiratory diseases, pulmonary damage and mortality among population (Mayer, 1999). Atmospheric chemical reactions transform primary pollutants into secondary air pollutants (Khoder, 2002). Atmospheric releases of acidic pollutants, in both gaseous and aerosol species, can cause adverse health effects and have the potential to cause other environmental damage (Gupta, 2003). Low emission source height and relatively high dry deposition velocity results in substantial fraction (20-40%) deposition near its source (Gupta, 2003). In India, pollution has become a great topic of debate at all levels and, especially, the air pollution because of the enhanced anthropogenic activities and global warming effects noticed in the recent past (Rao and Ansari, 2006).

For the purpose of emission inventory, the whole refinery is categorized into three main source-types, viz., point (stack), area (fugitive), and line (vehicular) sources. Specific information of various units, viz., production capacities, raw materials used, manufacturing processes, fuel consumption, stack emission characteristics, including stack heights and diameters, are obtained from the refinery personnel for identification of these sources. Emission sources in refineries differ greatly according to various processes, plants and units, and treatment procedures used. Various sources of pollutants are given in table 1. In this chapter, the methodology is discussed to undertake comprehensive green house gases (GHGs) emission inventory and air quality modeling for this industry.

**Table 1: Identification of Different Pollutants in Refinery**

| *Emissions* | *Sources* |
|---|---|
| Particulate matter | Cracking units using fluidized catalyst, steam boilers, furnaces, scrubbers, off-gas incinerators, product transfer. |
| Smoke | Inefficient combustion of heavy oils and tar, starting up and shutting down of heaters, flaring operations, product transfer. |
| $SO_4$ | Hydro cracking units, boilers, furnaces, flares, scrubbers, chemical processing units (burning of any sulphur containing fuel, subsequent burning of gases containing $H_2S$), product transfer. |
| $H_2S$ | May be emitted as above. |
| NOx | All combustion operations: Boilers, furnaces, flares and also catalyst regeneration, Product transfer. |
| CO | Catalyst regeneration, scrubbers, furnaces, off gas incinerators, etc., product transfer. |
| Hydrocarbons | Transfer and product loading operations, sampling valves, storage tanks, shut-downs and start-up operations, spills and leaks, oil separators, settling ponds. Relief valves, catalyst regeneration, pumps, compressors, cooling towers, vacuum towers, condensers, asphalt oxidizers, chemical processing units. |
| $NH_3$ and aldehydes | Catalyst regeneration unit. |
| Aromatics | Chemical processing units, asphalt oxidizers, Barometric condensers, sewers, vacuum towers, oil separators, etc. |

**Methods for Atmospheric Emission Inventory**

In the present context, unit stacks are referred to as point sources. There are about a total of 41 process/utility units as point sources in a refinery. Out of these, 28 are process units, 7 are utilities and others constitute 6 units. The fuels used in these units include FO and FG, while NG and coal/coke are also used in some utility units. In order to estimate the total emission load from different units (combustion sources), detailed stack monitoring should be done for criteria pollutants. The flue gas characterization should be undertaken as per standard such as IS: 5182 for $SO_2$ and $NO_X$. Fuel oil/ Fuel gas are the major fuels used in the units. The contribution of these fuels to various emissions should be estimated. Unit-wise process emissions have to be estimated to highlight the potential pollution sources. The emission factor and load have also to be estimated for each process stack.

Similarly emission load from area source fugitive emissions, which are mainly due to leakages and storage, should be estimated. Emission from these sources mainly depends on the type of storages, leakage source and its distribution. Hydrocarbons are the dominant pollutants being released from fugitive sources. An extensive area/fugitive emission survey has to be done for the refinery, and emission factor, and load from individual area source units has to be estimated for various conditions, such as storage tanks vented to the atmosphere, filling up of tanks with fixed roofs, temperatures changes, etc.

For line source survey, if the refinery under study is situated besides the main railway line and highway, and the refinery also has its own gantry for product loading, the line source inventory should take into account all these sources, and the load should be estimated. The GARB model for example, can be used for emission inventory. The format for data collection is presented in tables 2 to 6.

**Table 2 A: Instruments/Equipments used for Air Quality Monitoring and Analysis: CPCB Method**

| *S. N.* | *Instruments Name* | *Make* | *Parameters* |
|---|---|---|---|
| 1. | Spectrophotometer | Genesis 2, Photochem | $SO_2$ and $NO_x$ |
| 2. | Electronic Balance | Mettler toledo | SPM, RSPM |
| 3. | High & Low Volume Sampler | Envirotech | SPM, RSPM |
| 4. | Photovac 2020 | Perkin Elmer | HC in terms of isobutylene |

**Table 2 B: Analytical Techniques Used for Ambient Air Quality Monitoring**

| *S. N.* | *Parameter* | *Technique* | *Technical Protocol* | *Minimum Detectable Limit ($mvm^{-3}$)* |
|---|---|---|---|---|
| 1. | Suspended Participate matter | High Volume Sampler (Gravimetric Method) | IS-5182 &CPCB | 3.0 |
| 2. | Respirable Suspended Particulate matter | Respirable Dust Sampler (Gravimetric Method) | IS-5182 &CPCB | 3.0 |
| 3. | Sulphur dioxide | Modified west | IS-5182 & CPCB | 3.0 |
| 4. | Nitrogen oxide | Jacob & Hochheiser | IS-5182 & CPCB | 3.0 |
| 5. | Hydrocarbon | PID | USEPA | 5ppb |
| 6. | Wind Monitoring | APM 650 and SODAR | CPCB | |

**Table 3 : Emissions from Point Sources of Refinery**

| *S. N.* | *Category* |
|---|---|
| 1 | Process Units |
| 2 | Catalytic Cracking Units |
| 3 | Sulphur Recovery Unit |
| 4 | CO Boiler |
| 5 | Flare |
| 6 | Coke / Bitumen Blowing Unit |
| 7 | Gas Turbines |
| 8 | Thermal Power Plant* |

**Table 4: Emissions from Area Sources of Refinery**

| *S. N.* | *Category type* |
|---|---|
| 1 | Tank evaporation losses |
| 2 | Evaporation from API separator |
| 3 | Tank truck loading |
| 4 | Leakages from valves during blinding/ deblinding operations |
| 5 | Tank wagon Gantries |

**Table 5 : Emissions from Line Sources of Refinery**

| *Category type* | | |
|---|---|---|
| Tracks (product despatch/raw material) | | |
| In plant vehicles | | |
| | a. | Bus/Van |
| | b. | Jeep |
| | c. | Car |
| | d. | Forklift |
| | e. | JCB |
| | | Staff vehicles (Car) |
| Outsider/visitors' vehicles | | |
| Vehicles on highway road | | |

**Table 6: Total Emissions from all Sources of Refinery: Typical Format**

| | | Emissions ($kgh^{-1}$) | | | | | | |
|---|---|---|---|---|---|---|---|---|
| *S. N.* | *Category type* | *TSP* | $SO_2$ | $NO_X$ | *CO* | *NMHC* | *MHC* | *THC* |
| 1 | Point | | | | | | | |
| 2 | Area | | | | | | | |
| 3 | Line | | | | | | | |
| | Total | | | | | | | |

NMHC: Non Methane Hydrocarbon, MHC: Methane Hydrocarbons, THC: Total Hydrocarbon

**Methods for Atmospheric Emission Modeling**

Dispersion modeling is an acceptable technique for analyzing the impact of emissions from industrial sources. The United States Environmental Protection Agency (EPA) has developed guidelines on air quality models (EPA, 1986, 1993), and procedures on model evaluation (EPA, 1984, 1991). The total emissions from point, area and line sources can be estimated using CARB spread sheet model, and emission modeling can be based on critical winter season data of IMD, using USEPA prescribed ISCST-3 model, to arrive at the impact of refinery emissions on the air quality, and subsequently emission inventory refinement. The approach for modeling and AAQ is presented in Figs. 1 and 2.

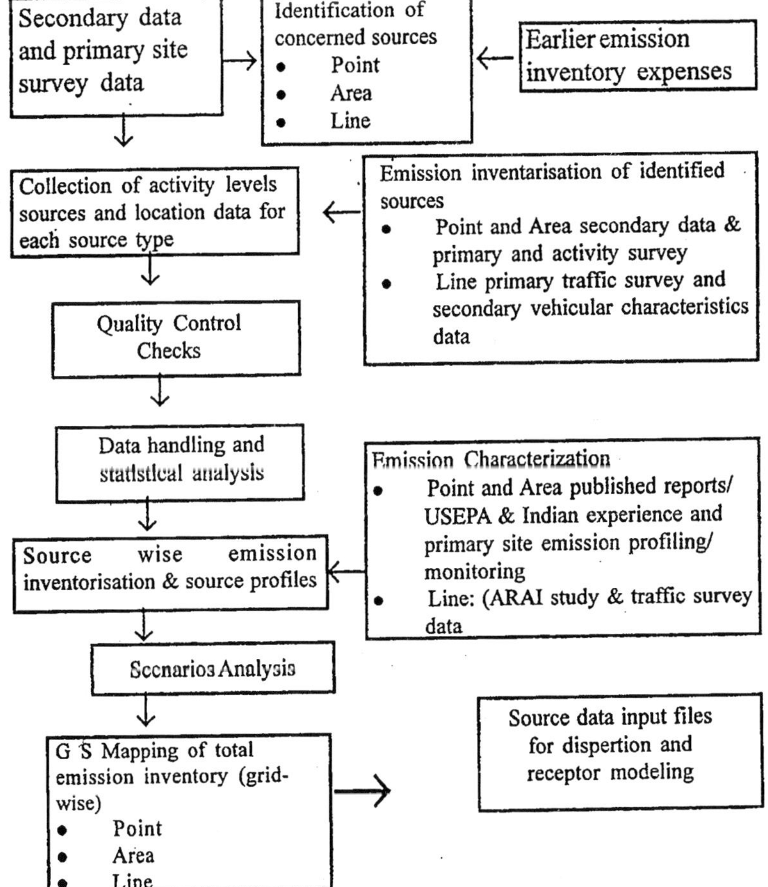


**Fig. 1: Approach for undertaking GHG Emission Inventory and Modeling**

| Particulars | Pollutants | | | | | | |
|---|---|---|---|---|---|---|---|
| | SPM | RPM | $SO_2$ | $NO_x$ | $PM_{2.5}$ | CO | VOC |
| Equipment | High Volume Sampler | Multi-speciation sampler/ Respirable dust sampler (For Delhi only) | Impingers attached to HVS | Impingers attached to HVS | $PM_{2.5}$ Analyzer | Automatic Analyzer | VOC Sampler |
| Measuring Principle | Aerodynamic Sampling followed by Gravimetric measurement | Gravimetric | Colorimetry | Colorimetry | Sampling by Impaction and measurement by Gravimetry | NDIR | GC-FID |
| Flow rate | 0.8-1.2m³/min⁻¹ | 0.8-1.2 m³/min | 0.5 lpm | 0.5 lpm | 1 m³/hr⁻¹ | 1 lpm | 0.5 lpm |
| Sampling period | 8/24 hrly | 8/24 hrly | 8/24 hrly | 8/24 hrly | 8/24 hrly | 4/24 hrly | One in a month |
| Sampling frequency | One month Continuous (3 seasons) | One month Continuous (3 seasons) | One month Continuous (3 seasons) | One month Continuous (3 seasons) | One month Continuous (3 seasons) | One month Continuous (3 seasons) | One week Continuous |
| Analytical Method | Gravimetric | Gravimetric | Improved West & Gaeke | Jacobs & Hochheiser modified | Gravimetric | NDR | GC-FD |
| Min. Detection limit | $1 \mu gm^{-3}$ | $1 \mu gm^{-3}$ | $0.04 \mu gm^{-3}$ | $0.03 \mu gm^{-3}$ | $0.5 \mu gm^{-3}$ | 0.04 ppm | $1 \mu gm^{-3}$ |
| Absorption wave length ($\lambda$ max) | - | - | 560nm | 550nm | - | | |
| Minimum reportable value | $10 \mu gm^{-3}$ | $10 \mu gm^{-3}$ | $3 \mu gm^{-3}$ | $9 \mu gm^{-3}$ | $5 \mu gm^{-3}$ | $4.67 \mu gm^{-3}$ | As specified in method |

No. of samples maximum in downwind direction, 24 hourly average for 3 months/season, twice a week Method IS: 5142

**Fig. 2: Approach for undertaking Ambient Air quality (AAQ) Monitoring around the refinery**

**ISCST3 Model Description**

Since the ISCST3 model (version 96113) is based on the previous ISCST2 and ISCST models, a brief history of these models precedes the ISCST3 model description. The ISCST2 (version 92062) model, or the revised Industrial Source Complex Short-Term Dispersion Model is a restructured, reprogrammed version of the previous ISCST model. This model (version 90436), in turn, was an extended version of the single source (CRSTER) model, written by the USEPA (USEPA, June 1986 and 1977).

The USEPA Model Change Bulletin MCB#1 (2 March 1992) for ISCST2 initiated the replacement of all previous versions of this program. At present, the revised version ISCST-3 is used by EPA. It contains advanced algorithms to determine plume rise and plume downwash where these factors are influenced by adjacent buildings. It also incorporates both the Schulman-Scire algorithm for shorter stacks and Huber-Snyder algorithm for stacks up to the good engineering practice stack height (e.g., typically to 65 m above ground level), when applicable. The ISCST3 model is, essentially, a refinement of ISCST2, with revised algorithms for area sources, and both dry and wet deposition modeling. A model algorithm for open pit sources has also been added to the ISCST3 model. The ISCST3 model is suitable for application to pollutant sources in the following types of studies :

- Stack design studies,
- Combustion source permit applications,
- Regulatory variance evaluations,
- Monitoring strategy evaluation for State implementation plans,
- Fuel (e.g., coal) conversion studies,
- Control technology evaluation,
- New source review (e.g., permitting of new emission sources), and
- Emission Inventory and Modeling studies, leading to prevention of significant air quality deterioration evaluations.

The model calculates ground level and elevated ambient concentrations or deposition from stack, volume and/or area sources. The steady-state Gaussian plume equation for a continuous source is used to calculate ground level concentrations for stack and volume sources. The area source equation in the ISCST3 model is based on the Gaussian point source plume formula, and is limited to rectangular areas where the ratio of length to width is a maximum of 10:1. This restriction can require extensive emission sources for a long narrow source such as a road.

Average concentrations or deposition rates may be calculated for 1-, 2-, 3-, 4-, 6-, 8-, 12- and 24-hour time periods. Annual average concentrations, or total deposition rates, may also be computed if the meteorological data used spans an entire year. Alternately, a monthly average concentration, or total deposition, may be computed when the meteorological data spans a monthly period. Concentrations or depositions may be calculated for all sources or for any one source.

**Modeling Protocol**

The following options can be chosen for the ISCST3 model run:

- Point, area and line source option,
- Discrete receptor option,
- No terrain consideration, if the vicinity is essentially flat with not more than a ten feet,
- No change in elevation within three kilometre radius,
- No building downwash, as all the stacks are sufficiently tall, and may not be affected by building downwash and aerodynamic wake effects.
- The combined impact of all stacks at each monitoring station, obtained by using the 'ALL' option and diagonal distances first calculated by using equation: $AB=\sqrt{(X1-X2)^2+(Y1-Y2)^2}$ (where AB = diagonal or direct distance in metres and XI, X2, Yl, and Y2 are coordinates in metres and individual impacts of each stack to be first obtained and then all individual impacts to particular monitoring stations to be added to get the combined impact.
- 24 hrs averaging periods,
- Hourly meteorological data to be obtained from IMD and/or Met station to be installed at the site,
- Anemometer height to be set at 10.05 m for model purpose,
- Urban mode can be selected,
- Default wind profile exponents and temperature gradient values can be chosen,
- Gradual plume rise option,
- Buoyancy induced dispersion option, and
- Calm processing option.

**Data Requirements**

The data requirements for modeling and analysis would consist of three important parts: the emission inventory, the meteorological data, and the monitoring data. A brief description of the data composition follows, and is given in table 7.

**Table 7: Checklist of Input Data (ISCST3 model)**

**General Information**

Submittal Date:

Facility Name:

Facility Location:

Modeler Name

**Air Dispersion Model Options**

a) Model selection:
- ISCST3

b) Dispersion Coefficients:
- Urban
- Rural
- Urban or Rural conditions can be determined through the use of an Area
- Land Use or Population Density analysis.

c) Coordinate System
- UTM Coordinates
- Local Coordinates
- Other

**Source Parameter Selection**

Summarize the reasoning for all emission rate and source parameter values, assumptions, locations, emission rates and release parameters for all point, area, and volume sources included in the modeling analysis.

**Building Downwash**

Is the stack(s) located within 5L of a structure that is at least 40% of the stack height (L is the length of the height or the

**Source Information**

a) Source Summary
- Summarize the locations, emission rates and release parameters for all point, area, and volume sources included in the modeling analysis. Information required is summarized in the tables below, each of which can be repeated as often as needed.

*b) Point Sources Summary*
- Name (ID) of the Source:
- Location:
- X(m):
- Y(m):
- Name of Pollutant Modeled Emission Rate [g/s]:

1)

2)

3)

Following detail information of stack has to be collected:
- Stack Height [m]:
- Stack Diameter [m]:
- Stack Exit Temperature [K]:
- Stack Exit Velocity [m/s]:
- Horizontal Stack

*b) Area Sources Summary*

*Table 7 : Contd.*

maximum projected building width for a structure).

- No.
- Yes. Perform a building downwash analysis using the current version of the Building Profile Input Program - PRIME (BPIP-PRIME) and include results in air dispersion modeling assessment.

**Scaled Plot Plan**

Provide a scaled plot, preferably in electronic format, displaying source, structure and related locations including:

- Emission Release Locations
- Buildings (On site and neighboring)
- Tanks (On site and neighboring)
- Property Boundary
- Model Receptor Locations
- Sensitive Receptors

**Receptor Information**

a) The following minimal receptor configuration must be met:

- Receptor definition must ensure coverage to capture the maximum pollutant concentration.
- Model runs with the following receptor densities would ensure that maximum ground level off property concentrations is captured.
- 20 m spacing within

- Source Name:

Location (Southwest Vertex):

- X(m):
- Y(m):
- Name of Pollutant Modeled Emission Rate [g/(s-m2)]

1)

2)

- Source Height [m]:
- Easterly Dimension [m]:
- Northerly Dimension [m]:
- Initial Vertical Dimension [m]:
- Angle From North [degrees]:

*c) Volume Sources Summary*

- Source Name:
- Location (Center of Source):
- X(m):
- Y(m):

Name of Pollutant Modeled Emission

Rate [g/s]

1) $SO_2$

2) NOx

3) CO

4) SPM

5) RPM

- Source Height (m):
- Initial Horizontal Dimension (m):

Initial Vertical Dimension (m):

**Terrain Conditions**

a) Does the modeled area contain elevated or complex terrain?

- No.
- Yes.

In both cases, provide a discussion on the approach used to determine terrain characteristics of the assessment area.

*Table 7 : Contd.*

| | |
|---|---|
| b) Fence line Receptors<br>Receptors must have not more than 50 metre spacing along property lines.<br>c) Sensitive Receptors<br>If applicable, provide a summary describing the location and nature of any nearby sensitive receptors (e.g., apartments, schools, etc.).<br>d) Capture of Maximum<br>Demonstrate that maximum has been reached and ensure the levels have dropped well below the standard and/or the guideline of the contaminants being studied. Describe the receptor coverage used to achieve this requirement.<br>**Check Box Meteorological Data**<br>• **Set**<br>• **Climatic**<br>• **Region**<br>• **Observation**<br>• **District/area**<br>**Results - Dispersion Model Predictions**<br>i) Model files - the following electronic model input and output files are to be provided:<br>• ISCST-3 Input and Output files<br>• ISCST-3 Plot files<br>ii) Meteorological Data - the following electronic meteorological data files must be provided: | b) Digital Terrain Data<br>• CDED 1-degree<br>• CDED 15-minute<br>• USGS 7.5-minute Ontario dataset<br>• Other<br>c) Elevation data import<br>• Describe the technique used to determine elevations of receptors and related model entities such as sources.<br>**Meteorological Data**<br>i) Whether the Preprocessed Regional Meteorological data used?<br>No.<br>If yes - Specify what data set was used from the typical table below and note the various requirements.<br>ii) Whether a Regional Meteorological Merge data files were used for large source category?<br>• No.<br>• If yes, specify the Meteorological Data Set Merge file used and summarize land characteristics specified in its processing.<br>iii) Were hourly surface data and upper air Regional Meteorological data files used?<br>• No.<br>• If yes, specify the Meteorological Data files used and summarize all steps and values used in processing these standard meteorological data files. |

*Table 7 : Contd.*

| | |
|---|---|
| • Preprocessed data files<br>• If files other than the Regional Preprocessed meteorological data files were used, you must include all meteorological data files as well as the AERMET input and output files.<br>iii) Terrain Data<br>• If elevated or complex terrain was considered, include the digital elevation terrain data files.<br>iv) Plots and Maps - include the following:<br>• Drawing/site plan with modeling coordinate system noted (digital format preferred.<br>• Plots displaying concentration/deposition results across study area.<br>v) Emission Summary<br>• An emission summary table should be provided. | iv) Whether, local meteorological data were used?<br>• No.<br>• If yes, specify the source, reliability, and representativeness of the local meteorological data as well as a discussion of data QA/QC and processing of data. State the time period of the measurements, wind direction dependent land use (if used), and any topographic or shoreline influences.<br>v) Wind Information - the following items should be provided and discussed in details wherever applicable<br>• Speed and direction distributions (wind roses)<br>• Topographic and/or obstruction impacts<br>• Data completeness<br>• Percentage of calms<br>vi) Temperature, clouds, and upper air data - the following items should be provided and discussed where applicable:<br>• Data completeness<br>• Mixing layer heights, diurnal and seasonal variations<br>vii) Turbulence - the following should be provided and discussed if site data is being used:<br>• Direct measurements - frequency distributions, diurnal and seasonal variations |

The emissions from various sources in the refinery can be estimated by field measurements and inventorisation, using CARB model. The emission inventory data requirements are stack details (source coordinates in grid, height, diameter, velocity, temperature and emission rate). The meteorological data requirements are wind profiles (hourly data on speed, direction, temperature, humidity for 3 months). The receptor data requirements are ground profiles (receptor coordinates in grid, height). All these data requirements can mostly be fulfilled by surveys and field measurements, and modeling can be undertaken.

## CONCLUSION

As per the provisions of Article 4 and 12.1 of the United Nations Framework Convention on Climate Change (UNFCCC), India has signed for the preparation of inventories of greenhouse gases (GHGs). This inventory, traditionally, involves gross energy based estimation and, hence, is subject to uncertainty. The Industrial emission inventory, therefore, has always been an area of research for a better estimation. Petroleum refinery, one of the major contributors to sulphur dioxide, carbon dioxide, oxides of nitrogen (NOx), methane and non-methane Volatile Organic Carbon (VOC) emissions, is a complex process and throws many challenges related to its emission inventory and management. The advantages of this study are that it will lead to emission inventory refinement, evaluation of air quality model, extension to other industries and would help in updating the regulatory database.

- Measurement of the emission coefficients, and preparation of emission inventory, for various refining activities.
- Air quality modeling, and its validation, using ambient measurements.
- Inventory refinement to reduce the gap between the modeling results and field measurements.

The advantages of this work is to provide insight to the contribution from different emission sources and facilitate in delineating effective management strategies, as also refining the existing emission regulation for the petroleum refinery.

## REFERENCES

Central Pollution Control Board (1995). *Emission Regulation:* 54: 770.

Code of Practice for Air Pollution Control in Petroleum Refineries (1981). *Indian Standards Institution,* New Delhi.

CONCAWE Report (1993). Survey on Oil Refineries Waste Disposal Methods Quantities and Cost, 1 : 38.

CONCAWE Report (1997). Regulatory Guidelines of Environmental Concern to the Oil Industry in Western Europe, 161 : 30.

*CONCAWE Review* (1995). 4(2): 6-7 and 10-11.

CONCAWE. Assessing Air Quality for European Auto / Oil Refinery: Initial Findings.

CPCB (1980-81). *Comprehensive Industry Document: Oil Refineries.* Central Pollution Control Board, New Delhi.

Crawford (1976). *Air Pollution Control Theory.* Mc-Graw Hill Inc., New York, 248.

Dobbins, Richard A. (2003). Environmental guidelines for Oil and Gas-Petroleum Refineries. *COFACE*, 187-298.

Hasan, M. Z. and Padma Rao (2001). Environmental Control in a Refinery: A Case Study. *Proc. RAWN Conference on Waste Minimization held at Banaras*, India, 24 : 123-129.

NEERI Report (1993-94). *Rapid Environmental Impact Assessment Studies of Gujarat Refinery*, India, submitted to IOCL, Gujarat Refinery, India.

NEERI Report (1998-99). *Rapid Environmental Impact Assessment Studies of Gujarat Refinery*, India.

NEERI Report (2000). Environmental Audit of Gujarat Refinery, Vadodara, submitted to IOCL, Gujarat Refinery.

NEERI Report (2001). *Environmental Audit of Gujarat Refinery, Vadodara,* submitted to IOCL, Gujarat Refinery.

NEERI Report (2002). *Environmental Audit of Gujarat Refinery, Vadodara,* submitted to IOCL, Gujarat Refinery.

Petroleum refinery fugitive emissions (http://www.baaqmd.gov/permit/handbook/petroleu.htm) pp. 1-7.

Rao, B.P.S. *et al.* (2005). Estimating Fugitive Emission Budget of Volatile Organic Carbon (VOC) in a Petroleum Refinery. *Bull. Environ. Contam. & Toxicol.*, Vol. 75, July 2005.

Rao, Padma, Ankam, S., Ansari M., Gavane, A.G. Kumar, A., Padit, V.I. and Nema, P. (2005). Monitoring of Hydrocarbon Emissions in a Petroleum Refinery. *Environmental Monitoring and Assessment*, 108: 123-132.

Rao, Padma S., Gavane, A.G., Ankam, S.S., Padit, V.I. and Nema, P. (2004). Performance Evaluation of Green belt: A Case Study. *Ecological Engineering*, 23: 77-84.

USEPA (2000). Petroleum refinery equipment leaks. *SIP-Ecology*, 173-490.

# Chapter 11

# Suitable Tree Species for Pollution Abatement

*Gayatri Verma*[1], *Alka Tomar*[2], *Yogender Singh*[3], and *Anjana Chowdhary*[4]

## INTRODUCTION

Environment has been defined "as the aggregate of all external conditions and influences affecting the life and development of organisms." Environmental pollution is one of the major problems, threatening the growing population of the world during the past decades due to rapid industrialization and urbanization. The environment in which we are living now is having three major constituents, viz., soil, water and air. An ideal environment is one in which one could have fresh air, pure water and noise free surrounding (Singh, 2003; Sivasamy and Srinivasan, 1997).

The most polluted city in the world is Tokyo. High levels of pollution have been found by NEERI, Nagpur (Saxena, 2003) in Mumbai, Delhi, Kanpur and Kolkata. The dust concentration in these cities is shown in table 1.

At the conference of International Environmental Organisation, held at Kota (Rajasthan, India) in the first week of October, 1981, Mr. Jun Ui of Tokyo University sounded a note of caution that developing countries like India should learn lessons from Japan where pollution had resulted in dreaded diseases like *Minamata, Itai - Itai and Kanemiyusho.* In our own country, incurable asbestosis (a scarring of lung) is developing speedily in Roro in Singhbhumi, Bihar (Indian Express, 1981).

[1]Department of Chemistry, BSA College, Mathura (UP) - 211004, India.

[2]CMS Environment (Research House), Community Centre, New Delhi-110017.

[3]Department of Biotechnology, College of Agriculture (Rajmata Vijayaraje Scindia Krishi Vishwavidyalaya), Ganj Basoda, Dist. Vidisha (MP), India.

[4]Department of Botany, Govt. Dungar College, Bikaner (Raj.) - 334001, India.

**Table 1: Suspended particulate matter**

| *S.N.* | *City* | *Suspended particulate matter* ($\mu gm^{-3}$) |
|---|---|---|
| 1 | Mumbai | 238 |
| 2 | Kolkata | 527 |
| 3 | Delhi | 700 |
| 4 | Kanpur | 488 |
| 5 | London | 224 |
| 6 | New York | 134 |

It is estimated that in Kolkata, atmosphere receives over 632 tonnes of pollutants every day.

It is further estimated that combustion of fuels produces $2.8 \times 10^8$ tonnes of carbon monoxide every year. The air-crafts with supersonic speed consume hundred of tonnes of fuel every hour. It has been worked out that each tonne of fuel, on combustion, produces 1.1 tonnes of steam, 1.1 tonnes of carbon dioxide, 0.1 tonne of carbon monoxide and 0.1 tonne of oxides of nitrogen, which get mixed up in the atmosphere. At the famous Heathrow airport in London alone, the heavy traffic of aircrafts annually released 10,000 tonnes of minute particles. It is only a pointer to the intensity of pollution. The problem of gaseous production due to combustion was known, centuries back and, in 1306, the British Parliament declared combustion of coal as illegal. In 1309, in Mexico, people were awarded death penalty for burning coal. Yet the use of coal has been steadily increasing. In 1870, only 2500 lakh tonnes of coal were taken out from mines while this figure increased to 28,000 lakh tonnes in 1970. Similarly, mineral oil was not in use till 1890 but by 1970, its use touched the quantity of 1200 crore barrels/year (Biswas and Dutta, 1994; Saxena, 2003).

The basis of our environment revolves round green leaf, and the forests provide the maximum leaf surface area. Plants are world's natural pollutants' sinks. Trees are known to fix and metabolize carbon dioxide and carbon monoxide, both photosynthetically and non-photosynthetically. According to a report, the global $CO_2$ production is estimated as $0.31 \times 10^{14}$ gallons per year. Trees are the most abundant, if not the most efficient, natural absorber of pollutants. A German Climatologist, Herman Flohn forecasts that due to persistent increase of

$CO_2$ in the atmosphere the earth, in a few years, will start experiencing the "green house effect" which, in its ultimate effect, will result in destruction of all species and leaving, ultimately, the earth as silent dead chunk of stone. This can be considerably neutralized by planting more trees and devoting more space for forests. Plants are of universal occurrence. They have adapted to survive in deserts, hills, water, cold, hot, and for that matter, in any type of climatic or soil conditions. Plants even appear on the sites of toxic industrial effluents. The National Botanical Research Institute, Lucknow has been successful in reclaiming sewage for human benefit by growing algae, *Spiruling plantensis,* which is a source of high protein super food (The Indian Express, 1981). Chlorella algae is also, similarly, useful.

In recent years, the utility of trees in environmental engineering, specially pollution problems has been slightly recognized. In the present article, various tree species suited for pollution control are described.

**Vehicular Pollution**

It is the primary cause of air pollution in the urban areas (60%), followed by industries (20-30%) and fossil fuel (Table 2).

**Table 2 : Vehicular population in major metropolitan cities in India**

| *Types of vehicles* | *Percentage* |
|---|---|
| 2/3 wheelers | 69 |
| Car, Jeep, Taxis | 14 |
| Bus, Trucks | 8 |
| Others | 9 |

About 1/5th of the vehicular population in India is concentrated in the major metropolitan cities. Among them, Delhi is the fourth most polluted city in the world, and the top most in India. About 2,000 metric tonnes of pollutants are emitted in the atmosphere by the vehicles every day (Sivasamy and Srinivasan, 1997).

The principal pollutants emitted by the vehicles are carbon monoxide (CO), hydrocarbons (HC), oxides of nitrogen (NO), and suspended particulate matter (SPM).

Central Pollution Control Board (CPCB) in Delhi indicates that vehicular activities contribute about 70% of the total quantity of emissions, while its impact in the region is about 98% of the total impact (Table 2).

**Table 3 : Ambient air quality standards in India**

| *Categories* | *SPM* | *SO* ($\mu gm^{-3}$) | *NO* |
|---|---|---|---|
| Industrial Region | 500 | 120 | 120 |
| Residential of Rural area | 200 | 80 | 80 |
| Sensitive areas like - hill stations, national parks, sanctuaries | 100 | 30 | 30 |

An annual average level of total suspended particulates (TSP) is five times higher than the standard set by the WHO in case of Delhi, Calcutta and Kanpur.

**Impact on Human Beings**

Several pollutants have their negative effects on human beings, either directly or indirectly (Jagdeesan, 2004). These are listed in table 4.

**Table 4: Various pollutants and their impacts on human beings**

| *Pollutant* | *Source* | *Effects* |
|---|---|---|
| Sulphur dioxide | Power house of petroleum industries | Suffocation, irritation of throat and eyes, respiratory diseases. |
| Nitrogen dioxide | Automobiles | Irritation, Bronchitis, Oedema of lungs |
| Hydrogen fluorides | Fertilizer and chemical industries | Irritation, disease of bone, mottling of teeth, respiratory diseases |
| Carcinogenic hydrocarbons | Chemical industries and automobiles | Cancer |
| Dust | Mines, quarries, ceramic factories, power stations | Respiratory diseases like silicosis, asbestosis, etc. |
| Noise (65decibels) | Automobiles, factories, loudspeakers, etc. | Insomnia, cardiac diseases, nervous disorders, etc. |
| (BOD* 3 $mgl^{-1}$) | Paper, sugar, textile & sugar mills, distilleries | Intestinal disorders, throat infections, tuberculosis etc. |

* Biological oxygen demand

Some aspects of pollution in relation to plants/trees are discussed as follows:

**1. Air Pollution Abatement**

The air pollution is caused mainly by the addition of obnoxious gases and particulate matter. Most of the gases are produced by combustion of fossil fuels in the automobiles industry, electricity generation and also for domestic consumption. The main air pollutants are (i) Sulphur dioxide (ii) carbon dioxide and carbon-monoxide (iii) ozone (iv) oxides of nitrogen (v) fluorides (vi) chlorine (vii) particulates and (viii) herbicides, etc. Of these, sulphur dioxide, ozone and herbicides cause health hazards.

Sulphur dioxide is produced mainly by burning of coal from industries that refine petroleum or metals, and from railways. Sulphur doxide emitted into atmosphere ultimately reaches down to earth, as acid rain. Such rains may have pH of 4.0 Even small quantities of hydrogen sulphide $H_2S$ (1 $mgm^{-3}$) produce offensive odour and cause irritation in nose. If concentration of $SO_2$ is 5 ppm, it leads to suffocation and there is risk of death. Lichens do take some gaseous $SO_2$. Sulphur dioxide concentration in the atmosphere in Delhi has been recorded at 0.233 ppm. Studies at National Environmental Engineering Research Institute, Nagpur have revealed that the contribution of Mathura refinery to the long term $SO_2$ concentration at Agra will be 1.2 $\mu gm^{-3}$, compared to the existing level of 15-20 $\mu gm^{-3}$ Short-term (1 hr.) peak concentration of 65$\mu gm^{-3}$ is expected under worst meteorological conditions in winter. Frequencies of such occurrences will be 2-4% in Agra region and 0 to 1% for Bharatpur region. Large scale plantations will help to keep this pollution load within safe limits. These figures are, however, low when compared to highly industrialized cities like London 250, Paris 110, Toronto 170, Milan 600 and Brussels 17 $\mu gm^{-3}$.

According to several estimates, the concentration of $CO_2$ has increased from 270 ppm to 320 ppm. The CO is highly poisonous to man as it lowers the oxygen carrying capacity of blood, and this deficiency of O, causes damage to living beings. It may cause giddiness and headache in less than an hour at a concentration of 100 ppm and such situations are common during traffic jams. Plants *(Ficus variegata, Coleus blumei,* etc.) utilize $CO_2$ by photosynthesis for their food and reduce this hazard (Shelke and Shelke, 2000).

Nitrogen oxides have a highly oxidizing action, killing plants and animals. Its greatest role is in the formation of smog (smoke and fog) in cold, humid conditions. Smog was first experienced in Los Angles in 1948 where it caused heavy damage. Recent studies have brought to light the fact that the buildings with airtight ceilings give rise to malady of 'Sick Building Disease'. This causes burning in the eyes, red spots on the skin and other allergic manifestations.

Agarwal (1991) has suggested growing of ornamental plants to mitigate certain gaseous pollutants, e.g., formaldehyde is absorbed by money plant, pathos, spider plant etc., and benzene is absorbed by *Gerbera, Daisier, Chrysanthemum.* Other plants with purifying qualities are bamboo, palms, peace lily, rubber plants, English ivy, and among the trees anjir *(Ficus carica),* bargad *(Ficus bengalensis),* pipal *(Ficus religiosa)* and mother-in-law's tongue (*Albizzia lebbek*), etc.

Dr. P.S. Panesar of IIT, Roorkee has reported that 10 tonnes lime kiln produces $CO_2$ to the extent of 11.86 tonnes day$^{-1}$, and if 170 oak trees *(Quercus rubra)* are grown, they will neutralize it. He has also worked out that 11 oak trees are good enough to neutralize the carbon dioxide produced by 1 tonne of cement production (in a 600 TPD cement plant, the $CO_2$ gas to be fixed is 459.64 tonnes).

The particulate matter arises from various industrial operations, mining, quarrying, grinding and polishing, saw mills, textiles and burning of the fuels. While the particles larger than $10\mu$ settle down rapidly, the smaller particles remain suspended for long. The particulate matter may also consist of various chemicals, pesticides and metallic compounds of tin, arsenic, lead, cadmium etc., which may cause tuberculosis and cancers. Air-borne herbicides used for control of weeds, such as 2,4-D ; 2,4-5 T are often volatilized, or particles are earned by the wind. The spray of pesticides on agricultural crops has ill-effects on the birds of 'Ranganthitoo' Bird Sanctuary (Mysore). In Delhi alone, 30 tonnes of hydrocarbons, 240 tonnes of $CO_2$, 2 tonnes of $SO_2$ and about 20 tonnes of oxides of $N_2$ are discharged everyday from motor vehicles. Thermal power stations of Delhi dump 6 tonnes of other toxic materials like arsenic, fluorides, etc.

Plants release oxygen through photosynthesis and purify air by introducing excess oxygen into the atmosphere, and by mixing fresh air into the polluted air. Some plants absorb air pollutants like hydrogen fluoride, SO, and $NO_2$. The pollutant least absorbed is CO. Species of Acer and Fagus absorb $SO_2$. A recent Russian study has established that a 500 m wide green area surrounding factories will reduce $SO_2$ concentration by 70% and nitric oxide concentration by 67 per cent (Saxena, 1981).

Trees increase wind turbulence, and this can be taken advantage of in dispersing off the air-borne pollutants to far off places. In respect of particulate air pollution like sand, dust, pollen, smoke etc., the surfaces of leaves, branches stems entrap them, which are, ultimately, washed off by rain. Around trees, there is more humidity and, thus, the suspended particles settle sooner on most of the trees in or surrounding the campus of factories, and also along unmetalled roads. Trees also mask fumes and disagreeable odours by replacing them with more pleasing foliage or floral odours, or even by actual absorption (Gupta, 1980).

Guidelines for plantings to reduce air pollution may be as follows:

1. Plantings should be perpendicular to the direction of prevailing winds,
2. Open and permeable plantings should be combined with dense plantings, and
3. Plantings should be concentric around the source of pollution.

**2. Acid Rain Abatement**

It may be recalled that a suffocating smog hovered over London in 1952 for many days, choking inhabitants of London and reducing visibility to nose length. Within two weeks 4000 deaths occurred. This dramatic incident hastened the attempts to clean-up the air pollution in large industrial areas. The simplest method adopted was to build tall smokestacks to send the emissions high into the atmosphere where they could be dispersed among the clouds and, thus, rendered rather harmless. The dwellers of London benefitted. What was ignored at that time, however, was that the act of sending emissions higher and far away gave birth to a new problem, the acid rain (Singh *et al.*, 2007).

Acid rain begins in the towering smoke-stacks of smelters and fossil fuelled power plants, oil refineries and other industrial funnels, and in the exhaust pipes of millions of overused cars that clog our major cities. Daily, thousands of tonnes of invisible gases - the oxides of sulphur and nitrogen-primarily stream into the atmosphere, following the winds on a journey that can last as long as several weeks (although on an average they remain aloft for 2 to 5 days), and are then taken thousands of kilometers from their origin. During this journey in the wind currents, chemical reactions take place which convert them into acid - causing sulphates and nitrates. What goes up, inevitably must come down. For sulphuric acid and nitric acid, the trip comes with every rainstorm and snowfall of the season.

Mixing with water vapors in the clouds they travel in, they eventually tumble down to earth, as acid rain and acid snow. Acid rain is capable of crippling and killing the environment they invade, already in Canada and USA thousands of rivers and lakes are 'dead' - no longer able to support fish or plant populations. In fact, it may not be an exaggeration if the life supporting rain may be rain of death in the next century. The Taj Mahal, Agra is already suffering from the bite of constant acid washings. The plants, through their myriad processes reduce the severity of such deadly acid rains (Anonymous, 1981).

### 3. Waste Water Pollution Abatement

The ever increasing growth of population and industrialization has resulted in greater demand for our water resources, and this, in its wake, is creating the problem of disposal of waste water. The quantity of sewage water in cities is rising day by day, as also effluents from industries like pulp and paper mills, textiles, refineries, etc. Although Central and State Water Pollution Prevention and Control Acts have been framed, yet our rivers are still the convenient disposal media for industrial waste, which is making our waters rather dead both for aquatic fauna and flora. We have heard of fires on the water in the Ganges in U.P.

The water discharged from mine areas could also be collected in pits or check dams and vegetation like *Ipomea aquatica* (nadi), *Trapa bispinosa* (Singhada) could be introduced. These can also be used for consumption by human beings. These water bodies could also be developed aesthetically by growing lilies and other aquatic plants in and around them. Some of the treatments given biologically to polluted water are NP/ Sewage Lagoons. Sewage lagoons or sewage filter beds are essentially oxidation ponds. The highly fertile effluent from sewage ponds, when diverted to these natural or man-made basins, permits vigorous growth of aquatic plants, which partially strip the harmful or odorous agents from the polluted water including, Ni, Cd, Hg, phenol, and potential carcinogens. The aquatic weeds promise to provide on indigenous source of fertilizers and soil conditioners. Common water plants are water hyacinth *(Eicharnia cressipes)*, bulrush *(Scirpus sp.)*, horn-wort *(Ceratophyllum demersum)*, hydrilla *(Hydrilla verticulata)* and duck weeds *(Lemna, Spirodella and Wolffui)*, and these are harvested to get fertilizers and, at the same time, removing some of the chemical impurities. Water hyacinth also reduces algal and faecal bacteria, surpended matter and odour causing compounds and, therefore, the resultant water is cleaner.

**Visual Impacts**

Development processes do cause long-term ugly visual impacts due to (i) surface excavations (ii) waste disposal dumping of solid wastes (iii) fixed plants (iv) mobile plants (mobile equipment and vehicles), (v) construction of sky-scraping buildings, and (vi) industrial complexes.

Well-designed landscape planning before commencing the activity is a must. Due to greed, need and misdeed, natural vegetation has been plundered and tree less moonscape on earth is appearing. A dead landscape is inert. Plants form walls and canopies in landscape and can be used to enclose, contain, enframe, link, enlarge, reduce, beautify the object. Plants provide living material for screening on undesirable or ugly spot, uncared for dilapidated building, junk yard, service areas, tailing dumps, solid waste mounds, large machinery, etc. Plants could be so grown as to seclude a small area from its surrounding and provide privacy. Besides this, proper placement of plants can make scenic views even more spectacular because of the wind effect they produce. Trees provide power to transform space of stones, bitumen and concrete into tapestries and mosaics of light and shadow. Trees are dynamic, giving different appearances in the changing seasons and throughout their life-span. Trees provide their own inherent beauty in all settings. Typical arrangement of grassy loans with shady trees indicates the appropriate setting for cultured life. Besides this, they provide direct benefit in the form of fuel, fodder and fruits and small timber etc., to the inhabitants and their cattle.

The choice of plants as a remedy for unpleasant visual intrusions introduced, will depend upon the micro-environment so wrought about.

**Forest Wealth of India**

Out of total geographical area of about 329 million hectares, forests cover about 75 million hectares, which works out to 19.5%, against the target of 33% for the plains and 66% for hilly regions. The gap should be filled with proper trees at the specified areas. Developed countries are having nearly 1.9 billion hectares of forest cover, while developing countries have 2.3 billion hectares (FAO, 2000).

**Trees**

Trees are woody plants, perennial in growth habit, with a spreading crown. They are grown for their economic or aesthetic value or both. They are grown for various purposes like species of shabby trees, avenue or roadside trees, screaming, fragrant flowers and checking air pollution. Some special characteristics of certain plant trees have been suggested to solve pollution problems:

1. Fleshy leaves that deaden round (e.g. Cacti, Euphorbia)
2. Branches that move and vibrate to absorb and mask sound (e.g. Putranjiva, Callistemone, Pongamia, Holoptelia, Leucaena, etc.)
3. Pubescene on leaves to entrap and hold dust particles (e.g. Guazama, Nyclanthes, Terminalia, etc.)
4. Stomata in the leaves to exchange gases (e.g. Acacia, Robinia, Tamarix, Zizyphus etc.)
5. Blossom and foliage that provide pleasant smell to mask unpleasant odour (e.g. Moringa, Cestrum, Eucalyptus, Michelia, Anona, Pterospermum, etc.).
6. Leaves and branches that slow wind (e.g. Acacia, Robinia, Tamarix, Zizyphus, etc.)
7. Leaves and branches that slow rain (e.g. *Ficus bengalensis,* Tamarindus, Delonix, Samania, Albizzia and Conifers, etc.)

**Selection Criteria**

Trees are to be selected based on the type of pollutants, their intensity, location, easy availability, and suitability of our climate. They have different morphological, physiological and biochemical mechanisms/ characters like branching habit, arrangement of leaves, size, shape, surfaces (smooth/hairy), presence or absence of trichomes, stomatal conductivity, proline content, ascorbic acid content, cationic peroxidase, and sulfite oxidase activities, etc. to trap or detoxify or reduce the pollutants. Some of the suitable tree species tolerant to different pollutants are given in table 5.

**Dust Pollution**

Trees having compact branching, closely arranged leaves, broad leaves of simple elliptical and hairy structures, shiny or waxy leaves and high proline content are suitable.

**Noise Pollution**

Trees having thick and fleshy leaves with petioles, flexible and capacity to withstand vibration are suitable. Heavier branches and trunks of the trees also deflect or refract the sound waves.

**Table 5: Trees tolerant to pollutants**

| Sulphur dioxide | Dust Pollution |
|---|---|
| *Albizia lebbeck* | *Alstoma macropnylia* |
| *Ailanthus excelsa* | *Cassia siamea* |
| *Alstonia scholaris* | *Dalbergia sissoo* |
| *Azadirachta indica* | *Ficus bengalensis* |
| *Ficus religiosa* | *F. infectoria* |
| *Lagestroemia flosreginae* | *Mangifera indica* |
| *Mimusops elangi* | *Peltophorum ferragineum* |
| *Polyathia longifolia* | *Polyathia longifolia* |
| *Teminalia arjuna* | *Shorea robusta* |
| *Acer platanoides* | *Syzygium cumini* |
| *Quercus palustris* | *Tactona grandis* |
| *Q. rubra* | *Alnus viridis* |
| **Ozone** | *Picea sp.* |
| *Acer plantanoides* | *Braya purpurascens* |
| *A. negunda* | *Salixplanifolia* |
| *Quercus rubra* | **Noise Pollution** |
| **Oxides of nitrogen** | *Alstoma scholaris* |
| *Fagus orientalis* | *Azadirachta indica* |
| *Quercus robur* | *Butea monosperma* |
| *Robinia pseudocacia* | *Erythrina variegata* |
| *Sambucusnigra* | *Grevillea robusta* |
| *Ulnus sp.* | *Pterospeiinum acerifoliu* |
| *Peroxy acetyl nitrate* | *Tamarindus indica* |
| *Acer platanoides* | *Terminalia arjuna* |
| *A. negunda* | *Acer negunda* |
| *Quercus palustris* | *Alnus indica* |
| *Q. rubra* | *Betula pendula* |
| **Lead** | *Cornus alba* |
| *Cassia siamea* | *Juniperus chinensis* |
| *Zizyphus mauritiana* | *Populus ferolunensis* |
| **Hydrogen fluorides** | *Syringa vulgaris* |
| *Ailanthus excelsa* | *Viburnum lantana* |
| *Juniperus sp.* | |

**Planting**

**a. For Air Pollution**

Jatropha *(Jatropha curcas)* can be planted by digging small pits of 45x45x45 cm at the required spacing for the production of biodiesel. Compared to diesel emissions, it reduces emissions of carbon monoxide by 44%, sulphates by 100%, unburnt hydrocarbon by 68%, particulate matter by 40%, polycyclic aromatic hydrocarbons (PAHs) by 80%, and the carcinogenic nitrated PAHs by 90% on an average. The overall ozone forming potential of biodiesel is 50% lesser than the diesel fuel (Kumar *et al.*, 2004)

**b. For Dust Pollution**

It is found that 8 m wide greenbelts between roads and buildings can reduce the dust fall by 2-3 times. Conifers can reduce the dust fall upto 42% in temperate urban areas.

**c. For Noise Pollution**

Different patterns of planting are adopted according to the speed of the vehicles. Greenbelts of (small and large) 18 to 30 m width, 15 to 27 m from traffic lanes, with central rows of atleast 13.5 m height are necessary for high speed vehicles. For moderate speed vehicles, green belts (small and large) of 6 to 15m width within the edge of belts from 6 to 15 from the center of the nearest traffic lane.

Shrubs of 1.8 to 4 m height should be planted next to traffic lane, followed by back up rows of trees 4.5 to 9 m tall (Saxena, 1993).

**Research Focus**

Research is under way at NBRI, Lucknow, NEERI, Nagpur and JNU, New Delhi. Research should also be focused on screening the indigenous trees tolerant / resistant to various pollutants and climatic requirements.

**Government Efforts/Initiatives**

In recognition of the felt need for environmental pollution control, various regulatory and promotional measures have been taken over the years in our country. These include -

- The Water (Prevention and Control of Pollution) Act, 1974, amended in 1988.
- The Air (Prevention and Control of Pollution) Act, 1981, amended in 1988.

- The Environment (Protection) Act, 1986.
- The Motor Vehicle Act, 1938, amended in 1988.
- An integrated Department of Environment, Forest and Wildlife in 1985.
- Apart from these, Noise (Pollution Control) Act is also on the anvil.
- Central Pollution Control Board (CPCB) and State Pollution Control Boards.
- Bureau of Indian Standards, ISI 3967,1994.

## CONCLUSION

Green plants are the largest oxygen producing factories of the world. They are the cheapest and most sustainable and perpetual modes of mitigating air, water, noise and visual pollutions. It is for the 'bio-technologists' to discover quick methods of producing desirable plants which can serve the purpose of biological control also, for e.g. *Acanthospermum* is suppressing another weed *Xanthium, Lantana* bug" controlled the menace of spread of *Lantana camara.* The tissue culture technique needs to be perfected and the propagation of plants through root suckers, branch cuttings and other vegetative organs should be improved.

While landscaping a city/town/industrial area, trees must be taken into consideration as a pollution sink, besides their bioaesthetical values. Annual need of oxygen for one person is met by 150m$^2$ of leaf surface, i.e., 30-40 m$^2$ of greeneries. Now, there is need to evolve a national level strategy to enact legislative measures for the development of green belts in and around polluted sites with suitable species to check the pollution effectively.

## REFERENCES

Agarwal, S. (1991). How to Reduce Pollution. *Ravivariya Parishista,* Rajasthan Patrika, dtd. 22.12.91.

Anonymous (1981). "Down-wind the acid rain-story". Environment ; Canada : Cat. No. En. 56/1981 E,ISBN-0-662-1 1421-23.

Biswas, D. and Dutta, S.A. (1994). Vehicular pollution : Combating the smog and noise in cities. *In: The Hindu Survey of the Environment,* 1994, pp. 41-45.

FAO (2000). The Global Forest Resources Assessment, 2000, Rome : FAO.

Gupta, R.K. (1980). *Plants for Environment Conservation* (Eds. B. Singh and M.P.Singh), Dehradun.

Jagadeesan, H.(2004). Pollution. Chapter 6, *In : Understanding Environment* (Eds. Kiran B. Chhokar, Mamata Pandya and Meena Raghuanthan). Sage Publications, New Delhi, pp. 136-152.

Kutnar, R.V., Ahlawat, S.P., Handa, A.K. and Gupta, V.K. (2004). *Jatropha Curcas :* The Fuel of the Future. *Employment News*, 29(34) : 1-2.

Saxena, V.S. (2003). Plants in Pollution Control. Chapter 12, In : *Current Environmental Issues* (Eds. B.B.S. Kapoor, Ahmed Ali, K.K. Singh and Chandrakanta) Madhu Publications, Bikaner, pp. 163-174.

Saxena, V.S. (1993). Forest Resources and their Management in Rajasthan, *In : Natural and Human Resources of Rajasthan* (Ed. T.S. Cliohan). Scientific Publishers, Jodhpur.

Saxena, V.S. (1999). Forest and Forestry in Rajasthan, In : *Arid Ecology* (Ed. S.C. Kalwar). Pointer Publication, Jaipur.

Saxena, V.S.(1981). Plants in the Improvement of Environment. Workshop on Modern Techniques of Site Identification, I.P.I. Dehradun, pp. 170-176.

Shelke, R.E. and Shelke, M.E. (2000). Environmental Pollution due to automatic exhaust emission and their adverse effect. *Environment & People*, 7(5): 19-20.

Singh, A.K., Tomar, Alka, Singh, K.K. and Phogat, V. (2007). Acid Rains and its harmful effects, Chapter 2, *In: Environmental Degradation and Protection,* Vol. I (Eds. K.K. Singh *et al.)* M.D. Publications, New Delhi, pp. 38-77.

Sivasamy, N. and Srinivasan, V.(1997). Environmental Pollution and its control by trees. *Employment News*, 21 (52) : 1-2.

The Indian Express, New Delhi, October 6, 1981, p. 4.

—oo—